KW-460-219

WITHDRAWN
FROM
UNIVERSITY OF PLYMOUTH
LIBRARY SERVICES

90 0667313 9

7 Day

University of Plymouth Library

Subject to status this item may be renewed
via your Voyager account

http://voyager.plymouth.ac.uk

Exeter tel: (01392) 475049
Exmouth tel: (01395) 255331
Plymouth tel: (01752) 232323

Agricultural Governance

Food security and sustainability are arguably the most important issues facing the agri-food sector at the beginning of a new millennium. In an era of globalization, where nation states appear to have a diminishing role in governing these matters, the existing and emerging power relations underpinning agri-food regulation demand renewed scholarly attention. Drawing upon the expertise of some of the most prominent writers in rural sociology, geography and anthropology, *Agricultural Governance* shows how globalization processes open up a new regulatory politics in which 'non-political' forms of governing play an increasingly influential role in shaping agricultural production and consumption.

With corporate actors assuming an important role in production and regulation, it is crucial that consideration is given to the political implications of these arrangements. This innovative book is the first of its kind to examine, in a critical and comprehensive manner, the characteristics of new forms of governing and regulation. Through detailed case studies in developed nations the contributors explore the relationship between globalization and new sites, spaces and agents of agricultural regulation. The essays demonstrate the political significance of regulatory mechanisms extending beyond the state and the consequences for the governing of the agri-food sector.

Vaughan Higgins is a lecturer in sociology at Monash University, Australia. His areas of research include agricultural regulation and the role of new technologies in rural governing. **Geoffrey Lawrence** is Professor of Sociology and Head of the School of Social Science at the University of Queensland, Brisbane. His work spans the areas of rural and regional development, globalization and localization, and social aspects of natural resource management.

Routledge advances in sociology

This series aims to present cutting-edge developments and debates within the field of sociology. It will provide a broad range of case studies and the latest theoretical perspectives, while covering a variety of topics, theories and issues from around the world. It is not confined to any particular school of thought.

Agricultural Governance

Globalization and the new politics of regulation

Edited by Vaughan Higgins and Geoffrey Lawrence

Routledge
Taylor & Francis Group

LONDON AND NEW YORK

UNIVERSITY OF PLYMOUTH

90 06673139

First published 2005
by Routledge
2 Park Square, Milton Park, Abingdon, Oxon OX14 4RN

Simultaneously published in the USA and Canada
by Routledge
270 Madison Ave, New York, NY 10016

Routledge is an imprint of the Taylor & Francis Group

© 2005 Vaughan Higgins and Geoffrey Lawrence selection and
editorial matter; the contributors their contributions

Typeset in Garamond by Wearset Ltd, Boldon, Tyne and Wear
Printed and bound in Great Britain by MPG Books Ltd, Bodmin

All rights reserved. No part of this book may be reprinted or
reproduced or utilized in any form or by any electronic, mechanical,
or other means, now known or hereafter invented, including
photocopying and recording, or in any information storage or
retrieval system, without permission in writing from the publishers.

British Library Cataloguing in Publication Data
A catalogue record for this book is available from the British Library

Library of Congress Cataloging in Publication Data
A catalog record for this book has been requested

ISBN 0-415-35229-0

Contents

Illustrations

Figures

Tables

Contributors

Carmen Bain is a graduate student in the Department of Sociology at Michigan State University and is interested in the social and ethical implications of changes to food and agricultural standards. She is currently researching the introduction of private sector agri-food standards in the Michigan blueberry industry.

Lawrence Busch is University Distinguished Professor of Sociology and Director of the Institute for Food and Agricultural Standards at Michigan State University. His interests include the strategic use of standards in the agro-food sector, higher education in agriculture, and the impact of the new biotechnologies on agriculture.

Hugh Campbell is Director of the Centre for the Study of Agriculture, Food and Environment (CSAFE) in the School of Social Sciences at Otago University, New Zealand. He has been the programme leader of a Public Good Science Fund programme: *Greening Food: Social and Industry Dynamics*, since 1995. He has conducted evaluations of industry development strategies deploying organic and IPM systems in kiwifruit, pipfruit, wine, honey and processed vegetable production. Since 2000, Hugh has been involved in New Zealand government activities surrounding the development of organic agriculture and was a member of the Government Working Group on Organics.

Lynda Cheshire is a lecturer in sociology in the School of Social Science at The University of Queensland. Her research interests include rural and regional development, governance and the state, community development and local responses to change and restructuring. She has published a number of book chapters and journal articles on her work and is currently completing a book entitled *Governing Development: Discourses and Practices of Rule in Australian Rural Policy*, which is due for publication, with Ashgate, in 2005.

Chris Cocklin is Professor of Geography and Director of the Monash Environment Institute at Monash University, Australia. His research interests include rural communities, rural land use, regulatory analysis and public policy.

B. James Deaton is a Research Associate with the Institute of Food and Agricultural Standards at Michigan State University. His recent research examines organic standards, third-party certification, food labelling, farmland preservation and the property value effect of hazardous waste sites.

Jacqui Dibden, Research Fellow with the Monash Regional Australia Project (MRAP), Monash University, Australia, has a background in social anthropology, history and community development. She has a particular interest in social change, and has carried out ethnographic research and social surveys in Indonesia and Australia, particularly in rural areas experiencing severe social dislocation. Since 1999, Jacqui has been employed as Project Co-ordinator and Research Fellow with MRAP. She has undertaken research on social capability of land managers, social impacts of rural restructuring, sustainability of rural towns, and (with Chris Cocklin) the impact of deregulation and other changes on the dairy industry in Australia.

Vaughan Higgins is a sociologist located at Monash University, Australia. His areas of research include agricultural regulation and the role of agri-environmental standards in rural governing. Vaughan's recent publications include *Constructing Reform: Economic Expertise and the Governing of Agricultural Change in Australia* (Nova Science, 2002); and *Environment, Society and Natural Resource Management: Theoretical Perspectives from Australasia and the Americas* (Edward Elgar, 2001).

Geoffrey Lawrence is Professor of Sociology and Head of the School of Social Science at The University of Queensland. His work spans the areas of rural and regional development; globalization and localization; and social aspects of natural resource management. He has been co-editor of the *International Journal of Sociology of Agriculture and Food*, and is currently on the editorial boards of the *Journal of Environmental Policy and Planning* and the *Journal of Sociology*. His most recent co-edited and co-authored books are *Recoding Nature: Critical Perspectives on Genetic Engineering* (UNSW Press, 2004); *Globalization, Localization and Sustainable Livelihoods* (Ashgate, 2003); and *A Future for Regional Australia: Escaping Global Misfortune* (Cambridge, 2001).

Richard Le Heron's long-standing agri-food research interests focus on transformations in agriculture and food under capitalist and other systems of organization. His book, *Globalised Agriculture. Political Choice* (Pergamon, 1993) explored the interdependencies between regulatory and accumulation regimes in agri-food production and consumption. Over the past decade he has examined issues of regulation and governance in the neo-liberalizing space of Australian and New Zealand agriculture and food, with particular emphasis on supply chain realignment. Most recently, his work has investigated the practices of business and governance that are constitutive of the globalizing food economy.

Les Levidow is a Research Fellow at the Open University, UK, where he has been studying the safety regulation and innovation of genetically modified crops. This research encompasses the European Union, the USA and their trade conflicts. These developments provide a case study of concepts such as regulatory science, sustainability, European integration, governance, transnational civil society and organizational learning. Research funding has come mainly from the European Commission and the UK's Economic and Social Research Council (ESRC). He has also been Managing Editor of *Science as Culture* since its inception in 1987, and of its predecessor, the *Radical Science Journal*.

Stewart Lockie is Director of the Centre for Social Science Research at Central Queensland University. His research interests lie in the sociology of food and agriculture, natural resource management and social impact assessment. Stewart is co-editor of a number of recent books including *Rurality Bites: The Social and Environmental Transformation of Rural Australia* (Sydney: Pluto Press); *Consuming Foods, Sustaining Environments* (Brisbane: Australian Academic Press); and *Environment, Society and Natural Resource Management* (Cheltenham: Edward Elgar).

Terry Marsden is Professor and Head of the Department of City and Regional Planning at Cardiff University. He is also established Chair of Environmental Policy and Planning, Director of the Environmental Planning Unit in the Department and co-director of the Economic and Social Research Councils' Research Centre for Business Relationships, Accountability, Sustainability and Society. His research interests include theoretical and empirical studies in rural development, agro-food studies, sustainable development and planning, policy development, and rural governance.

Philip McMichael is Professor and Chair of Development Sociology at Cornell University. He is a member of the FAO Scientific Advisory Committee on Food Security, and is on the Executive Board of the Global Studies Association. His works include *Settlers and the Agrarian Question: Foundations of Capitalism in Colonial Australia* (Cambridge University Press, 1984); *The Global Restructuring of Agro-Food Systems* (Cornell University Press, 1994, editor); *Food and Agrarian Systems in the World Economy* (1995, editor); and *Development and Social Change: A Global Perspective* (Pine Forge Press, 2004). His current research is on the politics of globalization, including a focus on global justice movements.

Mara Miele is a lecturer in agricultural economics at the University of Pisa and is also affiliated to the Centre for Business Relationships, Accountability, Sustainability and Society at Cardiff University. She is the author of many articles on organic food, rural sustainability and farm animal welfare, and of *Creating Sustainability: the Social Construction of The Market for Organic Products* (Circle for European Rural Studies, Wageningen University).

Jonathan Murdoch is Professor of Environmental Planning in the Department of City and Regional Planning at Cardiff University. He has published in leading sociology and geography journals on social and spatial theory, planning and governance, and quality and regulation in the food sector. He is also the author of a series of books on social and spatial change in the British countryside, including *Constructing the Countryside* in 1993 (UCL Press; with Terry Marsden, Philip Lowe, Richard Munton and Andrew Flynn), *Reconstituting Rurality* in 1994 (UCL Press; with Terry Marsden), and *The Differentiated Countryside* in 2003 (Routledge; with Philip Lowe, Neil Ward and Terry Marsden).

Emelie Peine is a Ph.D. candidate in the Department of Development Sociology at Cornell University. She is currently conducting research in Brazil for her dissertation, examining the politics of foreign investment in the soybean industry in the state of Mato Grosso.

Emma Roe graduated with a Ph.D. from the School of Geographical Sciences, University of Bristol, in 2002. The thesis studied embodied food consumption and non-human geographies. She was awarded an ESRC post-doctoral research fellowship in 2003 which she took up within the Geography Discipline at the Open University. Emma is currently working as a Research Associate on the EU Welfare Quality research project at the School of City and Regional Planning, Cardiff University.

Nell Salem is a Research Assistant in the Centre for Social Science Research and tutor in rural and environmental sociology at Central Queensland University. Her recent Honours thesis addressed the construction of self-identity and morality through vegetarian food choices.

Roberta Sonnino is a Research Associate in the School of City and Regional Planning, Cardiff University. She holds a Ph.D. in anthropology from the University of Kansas. Roberta's research interests include the theory and practice of sustainable development, tourism development, rural sustainability, economic and environmental anthropology, and local and alternative food chains.

Annie Stuart is a Research Associate in the Centre for the Study of Agriculture, Food and Environment (CSAFE) in the School of Social Sciences at Otago University, New Zealand.

Preface

This book explores the contemporary mechanisms and techniques that increasingly characterize the governing of agri-food industries within the developed world. Its basic premise is that globalization is creating new regulatory forms that are only now beginning to be understood by social scientists. While global forms of governing are seen – in their most obvious form – in the World Trade Organization, there are significant changes in how the conduct of actors is regulated within commodity chains and, at the local level, in the development of new institutional arrangements for food production and natural resource management. The social and institutional landscape of agri-food production has altered in fundamental ways over the past decade, with questions being asked by scholars from a variety of disciplines about the role of the state in regulating these arrangements. Significantly, the new forms of governing do not appear to be coalescing into a readily identifiable set of activities; rather, there is a fracturing of public and private arrangements and policies – creating spaces for the emergence of social contestation and protest.

While some excellent work exists by rural sociologists, geographers, social planners and political scientists on new forms of governing, there have been few attempts to link these or to reflect in a consistent manner on the broader regulatory implications of such changes. Our main motivation for this book was to make sense of the new forms of governing, and their impacts. A further motivation for this book was our concern that the emerging styles of governing, and the impacts of such regulation on the agri-food sector, were generally not being theorized in a coherent and critical manner. In fact, some very important areas remained undertheorized. Among these were: power relations in the regulation of the international food market; the place of the sub-national region in conforming to – as well as challenging – the activities of global capital; the extent to which the state continues to shape action while seemingly providing decision-making powers to local actors; the contradictory paths of development of genetically modified foods, and organic farming, under conditions of neoliberalism; the governing of both farmers and the rural environment; and the welfare of animals in the global marketplace. While it would be impossible to canvas all the issues surrounding contemporary agricultural governance, we have made every effort to provide insights into the areas mentioned above.

We thank all the contributors for their commitment to the production of this book, which commenced its life as a series of discussions between a number of Australian and New Zealand writers both inside, and outside, the Agri-food Research Network. The Network is a loose affiliation of geographers, sociologists and political scientists who meet annually to share ideas and information. It has grown from five members in 1993 to over sixty today, providing a forum for discussion of important developments in the agri-food arena.

We thank, in particular, our colleagues Dr Lynda Cheshire, Professor David Burch and Associate Professor Stewart Lockie for their insightful comments throughout the time during which the manuscript was being prepared. Our institutions – Monash University and The University of Queensland – provided us with the necessary time to complete the manuscript. We also acknowledge the efforts of Yeliz Ali and Joe Whiting from Routledge who provided editorial advice and assistance, and who enthusiastically supported the book. A number of research bodies provided funding for the work presented by contributors to this book. These bodies include the Australian Research Council, ESRC (UK), National Science Foundation (USA), and the European Commission.

Finally, our most important acknowledgement is to our partners – Melanie and Dimity – for their understanding, companionship and support. More than anyone else, they have understood our strong dedication to this project and have been prepared to make numerous sacrifices to allow us to bring the manuscript to fruition.

1 Introduction

Globalization and agricultural governance

Vaughan Higgins and Geoffrey Lawrence

The impact of globalization on agri-food industries has received significant attention from scholars in the social sciences over the past fifteen years. Drawing upon what has been termed 'agri-food globalization theory' (Buttel 2001), there is a broad recognition that the regulatory dynamics underpinning agriculture in Western nations have shifted dramatically. Where, previously, the nation-state exercised considerable control over the regulation of agriculture, the rise of transnational corporations (TNCs) in the agribusiness industries, and global governance agencies – such as the World Trade Organization (WTO), International Monetary Fund (IMF) and World Bank – has resulted in a reconfiguration of political power in which the state is no longer the predominant actor. Much excellent work has been conducted in the fields of agricultural sociology, geography and anthropology attempting to identify the key actors and processes behind this shift (see e.g. Bonanno *et al.* 1994; Burch *et al.* 1996; Goodman and Watts 1997; Le Heron 1993; McMichael 1994, 2004; Marsden *et al.* 1990). However, surprisingly little is known about *how* these actors and processes exercise an influence over production and consumption.

This book seeks to move beyond the existing literature on agri-food regulation by exploring the variety of techniques and practices that make possible agricultural regulation in a globalizing world. What is significant about these forms of governing is that they are not necessarily state-based, but comprise a mix of private and public regulation. Such techniques and practices, as a consequence, are frequently represented as non-political, and therefore as a more accurate and objective means for regulating agriculture than simply through the state. As the title of the book suggests, we believe that a conceptually coherent way of explaining and understanding these new arrangements is through the notion of governance. Not only does this concept demonstrate sensitivity to the diversity of actors involved in governing processes, but it also highlights the important role played by seemingly 'non-political' agents, using an array of new practices, in attempts to govern.

To appreciate the applicability of the concept of governance to studies of agri-food regulation, it is necessary first to review what is meant by the term. The discussion is necessarily brief, since reviews of the literature on governance may be found in Jessop (1995) and Stoker (1998), and applied to the rural sector in Goodwin (1998). The concept of governance is further developed in the works of scholars such as Kooiman (1993, 2003) and Rhodes (1997).

What is governance?

According to Pierre and Peters (2000), governance is a notoriously slippery term. It has become an umbrella concept for a wide variety of phenomena including policy networks, public management, coordination of sectors of the economy, public–private partnerships, corporate governance and 'good' governance as reflected in the objectives of global regulatory bodies such as the World Trade Organization (WTO) and World Bank (Pierre and Peters 2000: 14). Nevertheless, while the term *governance* is used in a variety of ways, and from a number of different theoretical perspectives, most scholars are united on at least one point: that it refers to a shift in regulatory arrangements where governing is not confined to a single domain such as, for example, the state. As Jessop (1995: 310) argues, 'the various approaches to governance share a rejection of the conceptual trinity of market-state-civil society which has tended to dominate mainstream analyses of modern societies'. Frequently viewed as distinct from one another, the insight of governance approaches is that they examine the alterations to the boundaries between these spheres of activity. Thus, where an analysis of governing may once have focused purely on the formal mechanisms of government within the state, it is now increasingly necessary to look at actors and mechanisms beyond the state. This shift in focus is summarized succinctly by Stoker (1998: 17) who notes that:

> governance refers to the development of governing styles in which the boundaries between and within the public and private sectors have become blurred. The essence of governance is its focus on governing mechanisms which do not rest on recourse to the authority and sanctions of government.

Thus govern*ment* is no longer as central to governing processes as it once was. The activity of governing is now shared between state-based institutions and agents that extend beyond the formal boundaries of government.

Changes in mechanisms of governing may be studied, Stoker (1998) argues, by adopting a governance perspective. Rather than developing a theory of governance, he outlines five propositions that provide a concise starting point 'for understanding changing processes of governing' (1998: 18). These propositions are:

1   Governance refers to a set of institutions and actors that are drawn from, but also beyond, the state.

2   Governance identifies the blurring of boundaries and responsibilities for tackling social and economic issues.
3   Governance identifies the power dependence in the relationships between institutions involved in collective action.
4   Governance is about autonomous self-governing networks of actors.
5   Governance recognizes the capacity to get things done which does not rest on the power of government to command or use its authority. It sees government as able to use new tools and techniques to steer and guide.

Stoker's five propositions might be considered a type of general sensitizing framework for this book that enables the range of actors and processes underpinning the governing of the agri-food sector to be identified. What is significant about these propositions is that they provide a useful point of reference in analysing the unique features of modern mechanisms of agricultural governing, as well as how they operate. As Goodwin (1998) argues, the adoption of a governance perspective raises important research questions and offers new conceptual possibilities in rural studies. We believe that the focus in this collection on governance as a perspective will do the same for the field of agricultural regulation and restructuring. To avoid any conceptual confusion we follow Kooiman (2003) in making a distinction between govern*ing* and govern*ance*. Where governing refers to the 'totality of interactions in which public and private actors participate', governance refers to theoretical conceptions of governing (Kooiman 2003: 4).

Changes in governing: the role of globalization

Globalization represents one of the key macro-social phenomena behind changes in governing mechanisms. Since globalization entered popular academic parlance in the 1980s, there has been substantial debate on whether this signals the end of the state as a form of sovereign authority, or the reconfiguration of state powers. While many academics have recently provided compelling evidence to suggest that the state does, in fact, remain a significant player in the rise of global relations (e.g. Hirst and Thompson 1999; Holton 1998; Scholte 2000; Weiss 1998, 2003), there is also widespread recognition that some of its main features have altered. According to Held (1991), the emergence of (1) a global economy; (2) transnational bodies; (3) international law; and, (4) hegemonic powers and power blocs contributes to changes in the role of the nation-state. These:

> combine to restrict the freedom of action of governments and states by blurring the boundaries of domestic politics; transforming the conditions of political decision-making; changing the institutional and organizational context of national polities; altering the legal framework

and administrative practices of governments; and obscuring the lines of responsibility and accountability of nation-states themselves.

(Held 1991: 157)

As state sovereignty is further restricted, new forms of governing emerge that operate at both a sub-state and a supra-state level. According to Jessop (1998: 32), governing occurs increasingly through heterarchic[1] means that have become 'more significant than markets or hierarchies for economic, political and social co-ordination'. The growing complexity and interconnectedness of the world economy that has occurred with globalization, as well as the associated undermining of state autonomy, has resulted in various attempts, at both global and local levels, 'to impose some structure and order through resort to heterarchic co-ordination' (Jessop 1998: 33).

In the globalization literature, most attention is frequently devoted to transworld governance by global regulatory bodies, such as the World Bank and IMF. As McMichael (2004) argues, these bodies both facilitate a 'globalisation project' and are important agents of global economic governance. Transworld governance institutions such as the WTO, the IMF and the Organization for Economic Cooperation and Development (OECD) have gained quite considerable regulatory power and competence, particularly in the economic policy surveillance of national governments (Scholte 2000: 148–149). Meanwhile, the management of global environmental degradation, regional conflict and human rights has also increasingly fallen under the ambit of transworld bodies such as the United Nations and special departments or programmes of the OECD, WTO and World Bank.

At the same time that transworld regulatory bodies have emerged, so too have a range of multilateral regional schemes. Dicken (2003: 147) identifies four types of regional trading blocs: (1) the free trade arena in which there is a strong preoccupation with the removal of trade restrictions between member states; (2) the customs union where there exists a common external trade policy towards non-members; (3) the common market in which there is free movement of factors of production between member states; and finally, (4) the economic union where there exists harmonization of economic policies under supranational control. The two most significant regional groups are the nations that have signed the North American Free Trade Agreement (NAFTA) (an example of a free trade region), and the European Union (EU) (which is the closest to full regional economic integration). In both cases, the regional economic blocs transcend the boundaries of individual nation-states and enable the freer flow of goods, services and information across national borders within these regions. As the situation in the EU makes clear, regionalism reconfigures national sovereignty. Thus the EU now has its own transnational sovereignty through trade liberalization, product standardization, a common currency for most member-states, and a single European market (see Holton 1998: 103). While regionalism is not simply a result of globalization, and may in fact be seen as part of a reaction against globalizing tendencies, it incorporates mechanisms of governing that extend beyond the state. For this reason, regionalism represents part of a broader shift in regulatory politics.

Globalization has also encouraged participation by private organizations in governing processes. This trend is referred to by Scholte (2000: 151) as a *privatization* of governing where there is increased 'scope by private-sector agencies to become involved in regulatory activities'. The breakdown of what Held (1991: 152) calls the post-war 'liberal consensus' limited the capacities for states to pursue Keynesian strategies of national economic management. As states progressively accepted economic interconnectedness, many – including Britain and the United States soon followed by Australia, New Zealand and Canada – adopted market-driven neoliberal policies of privatization and deregulation. Part of this shift involved private-sector agencies assuming an increased influence in processes of governing.

At a sub-state level privatization has occurred in two interrelated ways. First, there has been an increase in the use of public–private 'partnerships' as a seemingly more efficient means of governing (see Pierre 1998). Partnership arrangements involve a number of tasks previously undertaken by government that are 'contracted out' to private or quasi-autonomous sub-state agencies. Second, the rationality underpinning government intervention has shifted from a 'welfarist' focus on encouraging national growth 'through the promotion of social responsibility and the mutuality of social risk' (Rose and Miller 1992: 192), to one where the state seeks to facilitate the conditions for entrepreneurial self-governing. Thus the privatization of governing involves a focus on the individual rather than society *per se* as the legitimate site of regulation. Both partnerships and self-governance refigure the territory of governing since, as Rose (1999: 260) argues, 'the social logics of welfare bureaucracies are replaced by new logics of competition, market segmentation and service management'.

At a supra-state level, the privatization of governing is evident through the growing influence of TNCs and international non-government organizations (INGOs). Even though TNCs and INGOs may not always be involved directly in policy-making, their ability to exert pressure on both state and supra-state agencies makes them significant agents of governing (Held *et al.* 1999). Transnational corporations, for instance, are argued to have no allegiance to any one state. This is viewed as leading to a massive reorganization of national economies on a global scale in which large and highly mobile corporations are forcing nation-states to liberalize their trade and social policies in favour of market-driven neoliberal policies (McMichael 2004; Sklair 1995). If states resist this process their legitimacy is likely to be threatened as capital moves elsewhere in its search for optimum profitability. This means that states must create an environment conducive to capital accumulation, ensuring that trade and labour policies are geared towards the profit-making interest of TNCs. While states may have historically been characterized as 'centralizing agents', TNCs take a different role

6 Vaughan Higgins and Geoffrey Lawrence

as 'globalizing' agents given their part in binding together national economies on a global scale. INGOs, too, may be seen as globalizing agents. Including sociopolitical, human rights, professional and charitable bodies (see Holton 1998), the aims of INGOs are diverse, and not simply anti-globalization. For instance, some transborder associations have begun formulating their own regulatory instruments giving non-official bodies a significant role in governing (Scholte 2000: 154–156). These instruments cover such diverse areas as financial, food and environmental standards.

If, via globalization, new forms of governing are emerging beyond the traditional boundaries of government, how does this shape the politics of agricultural regulation? The approach taken in this book is to argue that the politics of agricultural regulation are altered in three important ways.

(Re)regulating spaces

For most of the middle part of the twentieth century, the state was the primary site through which 'national' agricultures were regulated. However, through globalization, state regulation is effectively restructured. While it is tempting to view these changes as part of a broader shift to deregulation and the rise of market rule over state rule, the contributions to Part I point to a greater complexity. Globalization, in fact, gives rise to new arrangements of regulatory space that are neither state- nor market-based. Multi-level partnerships, devolved decision-making and 'joined-up' institutional arrangements help to create a complex pattern of spatial reconfigurations (Karkkainen 2003). This point is developed in Chapter 2 where Emelie Peine and Philip McMichael explore the mechanisms of governing that make 'market rule' on a global scale possible. The chapter examines current forms of agricultural regulation in the global economy, arguing that the international food market is politically created and managed. While appearing to be of benefit to both North and South, the implementation of the 1995 WTO Agreement on Agriculture creates asymmetries that favour the profit-making interests of agribusiness and sustain the often substantial government support to farmers in developed nations. Thus such supranational institutional regulation of market relations politicizes the global economy, privileging northern states and affluent consumers at the expense of a majority of the world's population.

In Chapter 3 Lynda Cheshire and Geoffrey Lawrence explore the reshaping of the state as a space of governing. Traditional political-economic analyses of agri-food restructuring have tended to focus on how globalization places constraints upon the state's capacities to regulate the activities of TNCs, and prompts the establishment of new structures of governing that cross-cut and override national boundaries. According to Cheshire and Lawrence, this understanding of power is linear and 'top-down', neglecting the horizontal relationships established between state agencies and other actors, and the capacities of farmers to shape and transform power relations 'from below'. Drawing upon insights from governmentality and early actor-network theory, they argue that what is needed is a new way of conceptualizing agri-food governing that takes account of how power is exercised by local people and producers to reshape the state, its policies and its practices. Through the use of two examples, Cheshire and Lawrence assess how the Latourian notion of 'networks of association' might represent a more conceptually coherent way of examining how state agencies attempt to govern at the same time as taking into account the role of contestation in shaping programmes of rule. This approach, which focuses greater attention on the horizontal reconfiguration of state power, demonstrates how the state has not so much lost power, but governs increasingly through a loose network of state and non-state actors. In addition, the 'network' nature of power means that 'local' actors, such as farmers, have a more prominent place in governing processes: they are often able to counter-enrol state agencies to contest specific regulatory interventions and to advance their own goals and objectives in novel ways.

The significance of 'the local' is a theme taken up by Terry Marsden and Roberta Sonnino in Chapter 4. Focusing specifically on Europe, the authors explore the emergence of alternative forms of agri-food governing, in light of a reformed Common Agricultural Policy (CAP), and consider the implications for the development of a more sustainable rural development model. Of particular relevance for Marsden and Sonnino is the relocalization of food which stands in opposition to dominant modes of agri-food governing, but which also has recently begun to receive some EU support through the Rural Development Regulation (RDR). Drawing upon case studies from South-west England and Wales, the authors compare the different ways in which regional and localized food systems operate, the actors involved and the prospects for more sustainable production. Marsden and Sonnino argue that in both regional case studies an alternative paradigm focused on local and regional strategies for food production has begun to emerge. Through the forging of alliances between rural development, environmental and agricultural networks, there has been a re-evaluation of social, economic and environmental assets and hence a questioning of the dominant agri-industrial model. However, at the same time as an alternative approach to agri-food is emerging and receiving support, the development of a 'hygienic-bureaucratic' model – which focuses on accountability and the standardization of practices, and is led largely by corporate retailers – suggests the persistence of a neoliberal belief in 'the market' as the most sustainable means of governing agri-food. This theme is given sustained attention in Part II.

(De)politicizing practices

Global integration of the agri-food sector has given rise to new practices for governing food production that are not simply state-based. The increasing concern with food safety, quality, traceability and the overall sustainability of agri-food production has prompted concern that state regulatory measures alone are insufficient to deal adequately with the transborder flows that characterize contemporary food production and consumption. Equally, through the neoliberal belief that markets are more efficient, and less political, vehicles for regulation than are states, certain practices are introduced to ensure that production is oriented to 'market requirements'.

Chapter 5, by Carmen Bain, B. James Deaton and Lawrence Busch, shows how food standards represent one of the most significant emerging practices in the governing of food production. Prior to the formation of the WTO, standards for food safety, environmental quality, plant and animal health, and worker health and safety were largely the province of various government departments within nation-states. However, as Bain, Deaton and Busch argue, since the formation of the WTO, international standard-setting bodies, NGOs and the private sector have emerged as important agents, challenging the traditional forms of governing in the agricultural sector. For example, supermarket chains set standards for food safety that often exceed the standards promulgated by a country's government. Private standard-setting bodies play an increasingly significant role in agri-food regulation. Through the use of third-party systems of verification, standards are developed that enable harmonization of claims to food 'quality' and 'safety' along the entire supply chain. However, as Bain, Deaton and Busch point out, these types of structural change are far from benign in their effects. The burden of standards differs among market participants with actors such as developing countries and smaller farmers often not having the capacity to comply. In these cases, far from universally beneficial as is frequently the claim, standards expand the capacity of some participants, while limiting the capacity of others, to reshape social and economic relationships.

In Chapter 6 Hugh Campbell and Annie Stuart also focus on the role of standards in agri-food production. They use New Zealand as a case study to explore the significance of standards in the constitution of 'organic' commodities as governable objects. Campbell and Stuart note that over the past fifteen years, the New Zealand organic agriculture sector has – in a similar fashion to that which has occurred in other Western nations – been transformed from a social movement into an industry. While the initial theorization of power in agri-food analysis suggested a conflict between corporate capital and the organic social movement – centred in particular on issues of contracting, control of supply, control of price, and increasing commoditization – the eventual terrain of conflict that emerged has been over the processes for organic standards development. At a national level, a series of processes involving companies, certifiers, the organics movement and scientists negotiated each new revision of the standards. These efforts attempted to create an acceptable compromise between sustainability goals, commercial and trading needs, and the requirements of certification mechanisms of audit. However, as Campbell and Stuart argue, while harmonized organic standards are increasingly being negotiated to facilitate international trade in organic food, this has created several points of tension. EU and US organic standards are assuming prominence as the dominant standards for 'disciplining' the organic commodity, yet these have become abstracted from the localized sustainability issues within specific production spaces such as New Zealand. The tension between standardization and the local needs of growers diminishes the chances for organics to provide sustainable outcomes. Attempts to 're-localize' the audit, certification and renegotiation of organic standards have not yet proved to be successful.

Agricultural biotechnology represents another area where tension is evident between the complexities of local and regional sustainability, and attempts by state and corporate agents to promote standardized techniques for the governing of agri-food production. Les Levidow takes up this issue in Chapter 7 where conflicts over 'sustainability' are explored with specific reference to the GM debate. Rather than examining the role of standards, Levidow is interested in the divergent views of sustainability that arise in conflicts over GM crops, and the different priorities within each over what to sustain and how to best sustain it. Focusing on the European Union during the 1990s, he argues that early EU procedures favoured a view of the agri-environment as a homogeneous resource to be used in the interest of greater productivity – thereby enabling a broader range of potential effects from GM crops to be accommodated. Nevertheless, protest against agricultural biotechnologies from the late 1990s prompted a legitimacy crisis, giving way to more diverse national frameworks for regulation of GM crops. Such changes involved much greater scope for critical voices to be accommodated in regulatory and decision-making processes. Levidow argues that while this may be seen as a positive change, its actual effects need to be considered within broader limits of EU governing. Thus, even though a broader range of voices now participates in processes of governing, the methodological difficulties in operationalizing diverse environmental values, and the focus in EU innovation policy on 'economic competitiveness', has the potential to place limits on the promotion of alternatives.

In Chapter 8 Vaughan Higgins explores a somewhat different, but none the less important, practice of agricultural governing that is coming to be regarded as crucial in the management and planning practices of farmers: calculation. While calculation has long formed a central part of farm management, only more recently has it been linked to broader governmental rationalities seeking to improve the competitiveness and sustainability of national agricultures in an uncertain global market environment. Rather than a neutral means of responding to market pressures, calculation is viewed by Higgins as a key technology of modern agricultural governing that encourages farmers to reflect on their conduct in an advanced liberal way as 'calculative agents'. To provide evidence for this point, Higgins focuses on a training course seeking to build the planning capacities of farmers in the Australian dairy industry. He explores the technologies of calculation that are deployed through the course and the effects that these have on how farmers reflect on their planning practices. Higgins notes that the calculative technologies in the planning course render some features of farming practices more technically visible. This encourages farmers to focus upon those aspects of the farming enterprise able to be represented and manipulated statistically. As a consequence, the managerial conduct of farmers is 'configured' and 'responsibilized' according to the statistical representations made possible by the technologies. In this way, certain practices come to be constituted as more 'truthful' (and acceptable) than others in running a profitable dairy farm.

(Re)configuring objects and subjects of governing

The globalization of agri-food production contributes to the emergence of new sites of governing: the environment, consumers, animals and agri-food risks. In some countries, new hybrid organizational institutions (such as regional catchment management bodies) are being created that both constitute these sites as governable objects, and seek to address the range of social and natural resource management issues to which these sites give rise. They are emerging in the spaces that the state vacates as it is 'hollowed out'. Not surprisingly, therefore, the new institutions may be identified in regional socioeconomic development, in agri-food regulation and in animal welfare legislation. An important question is: How do such sites emerge and what are the institutions that enable them to become governable in the face of the (apparently growing) risks associated with food security and with environmental destruction? These questions are addressed in Part III.

In Chapter 9 Jacqui Dibden and Chris Cocklin examine sustainability and agri-environmental governing. Focusing specifically on Australia, they argue that in spite of a sustainability discourse having been present for at least two decades, there is a continuing incompatibility between current patterns of agricultural production, and of rural sustainability. Importantly, it would appear that none of the present mechanisms of governing the environment is challenging – in any fundamental manner – the productivist agricultural regime established in Australia. Drawing upon a case study of dairy deregulation, Dibden and Cocklin highlight the dilemma faced by many farmers – that of responding to price signals from an increasingly competitive international marketplace, yet obeying more stringent regulatory regimes, where compliance costs must usually be absorbed by producers. A number of confusing (contradictory) messages are sent to dairy producers, along with an array of often incompatible policy settings. Those in the dairy industry have experienced the economic consequences of deregulation – falling prices, increased feed costs, reduced access to resources such as irrigated water and so on. Leaving the industry or increasing the size of the milking herd have been the main options. Those staying in the industry have been placing major strains on the environment, literally working their farms (and themselves) harder to make a living from dairying. Against the background of drought, the attempt by the state to 'force' higher productivity from dairy farmers in the context of the re-regulation of the environment exposes a fundamental incompatibility between broader neoliberal settings for agriculture, and local demands for environmental security.

While Dibden and Cocklin examine agri-environmental governing from a production perspective, Stewart Lockie and Nell Salem (Chapter 10) focus on 'the environment' from a different angle – that of consumption. Lockie and Salem investigate the strategies used to enrol people as consumers in networks of commodity production and consumption involving genetically modified (GM) and organic foods. Building on Lockie's (2002) earlier work on consumption, the chapter is concerned particularly with the 'technologies of the self' that are deployed, in the absence of the direct regulation of consumption practices, to influence the ways in which potential consumers are likely to understand GM and organic foods and their own relationship to them. For Lockie and Salem, media discourses, marketing and advertising, labelling laws and so on may each be seen to embody competing claims to expertise and knowledge that attempt to link the strategic goals of GM and organic proponents with consumers' self-identities, beliefs and practices. While attempts to shape consumption activities are unique neither to GM nor to organic foods, these serve, according to the authors, as particularly useful examples given the challenges which both face in mobilizing consumers and their diametrically opposed approaches to the regulation of production and product labelling. In addition to the governmentality perspective, Lockie and Salem also draw upon arguments within the sociology of science and technology regarding the need to examine the role of non-humans in the networks of the social. Applying these theoretical insights, they note that attempts to enrol consumers are a contingent matter hinging on their understanding of how successful, or otherwise, these networks have already been in enrolling or excluding other organisms ranging from the creations of genetic engineering to pests and pathogens.

In Chapter 11 Mara Miele, Jonathan Murdoch and Emma Roe focus on issues surrounding the governing of one variety of these non-human 'Others': animals. According to the authors, animal welfare has become of increasing concern to many producers and consumers of food; thus governments have been forced to recognize that animals are more than just 'machines' but may be living, sentient beings in need of protection against gross exploitation. Yet, at the same time as public concern about the conditions of animals has been growing, the use of animals in food production has been accelerating so that currently billions of animals are consumed annually around the world. For Miele, Murdoch and Roe this acceleration has given the issue of animal welfare even more significance, forcing some governments to act. Focusing on the policy environment in the United Kingdom and, more broadly, in the European Union, they argue that an increasing recognition of animals as sentient beings may be contributing to a new governmentality of animal welfare with its own rationalities and technologies. However, while animal welfare is moving closer to the centre of policy, problems remain concerning differing objectives and standards among the various scales of government. These ambiguities highlight more general cultural ambivalences over the status of animals.

Part III concludes in Chapter 12 with Richard Le Heron examining the role of risk in agri-food governing, and particularly the rise of a 'culture of riskification'. Over the past decade agri-food risks have been increasingly identified as objects of governing requiring new styles of management through modification of existing, and the adoption of new, governmental strategies. With a specific focus on biosecurity in New Zealand, Le Heron explores from a post-structural political economy perspective the emergence of new practices of conception, calculation and competence in agriculture, business and government that underpin the multiple discourses of risk now associated with the agri-food sphere. He situates the construction of 'risk' as a category in the context of agri-food restructuring, the rise of neoliberal political rationalities and spatial imaginaries. Le Heron examines how a new generation of expertise has been mobilized to deal with local and international crises such as food scares, biosecurity breaches, market collapse and environmental imaging. The chapter argues that the commodification of risk, through discourses of 'international competitiveness', is a qualitatively different framing of the relations between agri-food producers and consumers. In this context, new constructions of individual behaviour and social outcomes are constitutive of a deepening of accumulation processes, which pose further political challenges for the state and citizenry.

Conclusion

Throughout the world there has been a subtle but perceptible shift in the ways individuals, communities, natural resources and 'spaces' are being governed. As globalization has proceeded, new groups of political actors, quasi-government authorities, private organizations and regional entities have emerged both to contest current forms of governing, as well as to provide concrete alternatives. These cut across, and sometimes undermine, older forms of government. The more fluid arrangements that have subsequently arisen are part of a polyarchic 'mixed actor' system in which power is diffused rather than centralized (Held *et al.* 1999). In terms of agriculture, the previous role of many nation-states in protecting, subsidizing, and in various other ways directly and indirectly supporting rural producers has – under neoliberal rationalities – been altered in a manner that renders the political authority of the state much more limited.

Private regulations (devised in some circumstances without reference to state policies) have seen some commodity groups increase the share of domestic and world markets – a certain sign that market standards, taken on board by commercial operators, are part of the drive to change the current system of agri-food governing. New strategic arrangements between producer groups represent another way that the 'local' is dealing with the 'global'. An example here is the way in which some farmers have formed alliances across time and space (that is, beyond the nation-state) to guarantee supply to the large supermarket chains, whose power has grown enormously over the past decade (see Burch and Lawrence 2004). Yet another example is that of the private regulation of the public sphere, in which food retailers seek to surpass public food standards in their attempts to gain legitimacy from the public as the defenders of consumers' interests. The emergence of EUREP-GAP is a clear indication that changes at the global level are fostering the development of private regulatory entities that hold great sway over producers, literally forcing them to abide by the new rules of the food retail sector, or lose market share. Such re-regulation is totally consistent with WTO measures to expand the 'free trade' agenda.

Finally, we must recognize that regulation by, and other activities of, private capital is not going uncontested. Various NGOs and activist groups are targeting corporate capital, placing a great deal of pressure on firms to ban genetically modified ingredients in foods, offer more health choices to the consumers of fast foods, and to source foods from sustainable production systems, and from areas around the world where social justice is a guiding principle in the hiring of agricultural workers. We are in agreement with Busch and Bain (2004) who argue that such initiatives as private labels, direct contracting, third-party certification schemes, and a host of other privately directed activities are leading to the establishment of new rules, institutions, networks and conventions, and that the latter are part of a new scheme through which the agri-food system is governed.

The global reshaping of the agri-food sector forces us to examine questions of governing and governance. The task has only just begun. Indeed, this book does not attempt to devise a theory of agricultural governance. Rather, it seeks to use the notion of governance as an heuristic starting point to investigate the multiple forms of governing that increasingly make up the regulation of the agri-food sector in Western nations. The focus is upon different modes and practices of governing, how they have emerged, the forms of knowledge on which they make their claims to truth, and the politics of regulation to which they give rise. We believe that the analyses of governing assembled in this book provide a crucial starting point in rethinking the conceptualization of regulation in rural (and in particular agri-food) change.

Note

1 Heterarchy refers to self-organized, as opposed to hierarchical 'top-down', forms of governing. According to Jessop (1998: 29) its forms include 'self-organizing interpersonal networks, negotiated inter-organizational co-ordination, and decentred, context-mediated systemic steering'.

References

Bonanno, A., Busch, L., Friedland, W.H., Gouveia, L. and Mingione, E. (eds) (1994) *From Columbus to ConAgra: The Globalization of Agriculture and Food*, Lawrence: University Press of Kansas.

Burch, D. and Lawrence, G. (2004) 'Supermarket own brands, supply chains and the changing agri-food system: the UK experience', Paper presented at the Annual Meeting of the Agri-food Research Network, Australian National University, Canberra, 9–11 June.

Burch, D., Rickson, R. and Lawrence, G. (eds) (1996) *Globalization and Agri-food Restructuring: Perspectives from the Australasia Region*, Aldershot: Ashgate.

Busch, L. and Bain, C. (2004) 'New, improved? The transformation of the global agrifood system', *Rural Sociology*, 69, 3: 321–346.

Buttel, F.H. (2001) 'Some reflections on late twentieth century agrarian political economy', *Sociologia Ruralis*, 41, 2: 165–181.

Dicken, P. (2003) *Global Shift: Reshaping the Global Economic Map in the 21st Century* (5th edn), London: Sage.

Goodman, D. and Watts, M. (eds) (1997) *Globalising Food: Agrarian Questions and Global Restructuring*, London: Routledge.

Goodwin, M. (1998) 'The governance of rural areas: some emerging research issues and agendas', *Journal of Rural Studies*, 14, 1: 5–12.

Held, D. (1991) 'Democracy, the nation-state and the global system', *Economy and Society*, 20, 2: 138–172.

Held, D., McGrew, A., Goldblatt, D. and Perraton, J. (1999) *Global Transformations: Politics, Economics and Culture*, Cambridge: Polity Press.

Hirst, P. and Thompson, G. (1999) *Globalization in Question: The International Economy and the Possibilities of Governance* (2nd edn), Cambridge: Polity Press.

Holton, R.J. (1998) *Globalization and the Nation-State*, Basingstoke: Palgrave.

Jessop, B. (1995) 'The regulation approach, governance and post-Fordism: alternative perspectives on economic and political change?', *Economy and Society*, 24, 3: 307–333.

Jessop, B. (1998) 'The rise of governance and the risks of failure: the case of economic development', *International Social Science Journal*, 155: 29–45.

Karkkainen, B. (2003) 'Toward ecologically sustainable democracy?', in A. Fung and E. Wright (eds) *Deepening Democracy: Institutional Innovations in Empowered Participatory Governance*, London: Verso.

Kooiman, J. (ed.) (1993) *Modern Governance: Government–Society Interactions*, London: Sage.

Kooiman, J. (2003) *Governing as Governance*, London: Sage.

Le Heron, R. (1993) *Globalized Agriculture: Political Choice*, Oxford: Pergamon Press.

Lockie, S. (2002) '"The invisible mouth": mobilizing "the consumer" in food production–consumption networks', *Sociologia Ruralis*, 42, 4: 278–294.

McMichael, P. (ed.) (1994) *The Global Restructuring of Agro-food Systems*, Ithaca, NY: Cornell University Press.

McMichael, P. (2004) *Development and Social Change: A Global Perspective* (3rd edn), Thousand Oaks, CA: Pine Forge Press.

Marsden, T., Lowe, P. and Whatmore, S. (eds) (1990) *Rural Restructuring: Global Processes and their Responses*, London: David Fulton.

Pierre, J. (ed.) (1998) *Partnerships in Urban Governance*, Basingstoke: Macmillan.

Pierre, J. and Peters, B.G. (2000) *Governance, Politics and the State*, Basingstoke: Macmillan.

Rhodes, R.A.W. (1997) *Understanding Governance: Policy Networks, Governance, Reflexivity and Accountability*, Buckingham: Open University Press.

Rose, N. (1999) *Powers of Freedom: Reframing Political Thought*, Cambridge: Cambridge University Press.

Rose, N. and Miller, P. (1992) 'Political power beyond the state: problematics of government', *British Journal of Sociology*, 42, 2: 173–205.

Scholte, J.A. (2000) *Globalization: A Critical Introduction*, Basingstoke: Palgrave.

Sklair, L. (1995) *Sociology of the Global System* (2nd edn), Hemel Hempstead: Harvester Wheatsheaf.

Stoker, G. (1998) 'Governance as theory: five propositions', *International Social Science Journal*, 155: 17–28.

Weiss, L. (1998) *The Myth of the Powerless State: Governing the Economy in a Global Era*, Cambridge: Polity Press.

Weiss, L. (2003) *States in the Global Economy: Bringing Domestic Institutions Back In*, Cambridge: Cambridge University Press.

Part I
(Re)regulating spaces

2 Globalization and global governance

Emelie Peine and Philip McMichael

Introduction

The appearance of the term 'governance' coincides with the so-called era of globalization. The association of governance with globalization is twofold: first, governing is increasingly the domain of non-state organizations (whether multilateral institutions or non-governmental or corporate organizations); and second, governance is a euphemism for private power (exercised through the market). Indeed, globalization has become, discursively, a form of governing itself. The historical context for this extended meaning of governance is the deterritorialization of space, through the deepening of market relations. This is accomplished along three related dimensions: (1) the extension of commodity circuits, from seeds to services; (2) the centralization of capital in transnational corporate organizations (whether firms or strategic alliances among firms); and (3) the privatization of public institutions, associated with the neoliberal prescription for expanding markets and shrinking states. In this scenario, globalization is understood as the reduction of market friction (i.e. state interference), in order to realize the efficiency of market-based resource allocation. And governance is understood, primarily, as the management of market relations across the whole gamut of social and environmental arenas (from health care, through financial services to pollution permits).

Globalization involves a discursive reordering of the world, a representation of the market as a rational instrument of human progress and global development: in short, 'market rule' (cf. Arrighi 1982). Our goal is to examine how market rule is elaborated as a governing discourse. We use agricultural regulation as our case study – in particular the implementation of the World Trade Organization's 1995 Agreement on Agriculture. While the WTO is often regarded as a global governing body, the project to liberalize agriculture is a watershed example of the dilemmas and contradictions of globalization and governing. Agriculture is associated, and often represented, as a national resource, but it has been increasingly constructed as a global economic value, via the discourse of market rule and comparative advantage. National agricultures, therefore, are brought into competitive relation to one another via market rule, which, we argue, is anything but rational in its substantive social and cultural consequences.

Globalization and governance

The case of the WTO focuses on the symbolic shift from GATT government to new forms of governing through the WTO. The General Agreement on Tariffs and Trade (1947) rested on the principle of national sovereignty informing international agreements in the twentieth century. This was the century in which the Westphalian state-centric notion of sovereignty produced the League of Nations and, later, the United Nations. In the post-Second World War world, the 1990s transformation of the GATT into the WTO coincided with global market integration, expressing a movement from (national) government to (international) governing. The WTO essentially institutionalized the principle of market freedoms (trade, capital movement and equal treatment of foreign investors), requiring all member states to submit to a dispute resolution mechanism, by which panel rulings against protectionist measures are automatic and binding, requiring total consensus to reverse a ruling.

In many ways, the WTO dispute settlement mechanism is a concrete expression of the arrival of new forms of governing for the twenty-first century. That is, states internalize market rule by authoring free trade agreements that privilege corporations over citizens, and by applying private measures to the delivery of public goods. In each sense, states govern in the interests of private property. Whether and to what extent this introduces a different order of social and political relations, where multiple private enterprises replace singular public bureaucracies, where horizontal relations replace vertical relations, is a matter of debate.

To the extent that the experience of globalization is the elevation of flows (market exchanges) over spaces (state organizations), a new literature accounting for the elaboration of governing, beyond government, has appeared. Ruggie notes: 'the process of international governance has come to be associated with the concept of international regimes, occupying an ontological space somewhere between the level of formal organizations, on the one hand, and systemic factors, on the other' (1998: 89). The new literature, wavering between rather abstract definitions of governing and quite concrete applications, includes a special issue of the *International Social Science Journal* (1998) on governance. Here, Stoker suggests, 'governance is ultimately concerned with creating the conditions for ordered rule and collective action. The outputs of governance are not therefore different from those of government. It is rather a matter of a difference in processes' (1998: 17). Kazancigil argues that the governance model promises 'a greater capacity to cope with policy-making issues in increasingly differentiated modern societies, where the various social sub-systems and networks have become more autonomous' (1998: 70).

References to autonomous sub-systems, networks and difference in processes imply transcendence of hierarchy, and the significance of mutual interaction among a multiplicity of governing actors (cf. Kooiman and Van Vliet 1993: 64). This means essentially that government of a single territory by a central state is now complicated and/or transcended by governing of inter-societal and inter-state relations and transnational processes. Jessop views the 'growing fascination with governance mechanisms as a solution to market and/or state failure', noting that one key genealogical derivation of governance is the Enlightenment principle of heterarchy, or self-organization (in this case of inter-organizational relations), claiming, 'This form of governance involves the coordination of differentiated institutional orders or functional systems ... each of which has its own complex operational logic such that it is impossible to exercise effective overall control of its development from outside that system' (1998: 30). In our view, the use of the state/market binary, and the suggestion of autonomy, belies the internalization of the market principle, which, while amplifying market anarchy, reinterprets public goals in private terms and enhances corporate power. It is symptomatic of the tendency, particularly among international relations scholars, to treat globalization as an 'independent variable' (cf. Prakash and Hart 2000).

At the global level, the shift from government to new forms of governing may be represented ideal-typically as a movement from hierarchical power in the states system, to negotiated power in a context in which national sovereignty yields to alternative scales and dimensions of political coordination, affecting states and yet transcending them at the same time. While states remain the key guardians of territorial sovereignty, international flows sometimes override national spaces, and governance is the coordination of that process and its consequences. As de Senarclens observes: 'It is indisputable, nonetheless, that globalization weakens the ability of states to defend a type of economic and social regulation that had been linked to the defence of the modern idea of citizenship' (1998: 102). This apparent weakening of the *nation* state by globalization lends credibility to the notion of governance as a significant concept for explaining political and historical developments in governing.

Ultimately, our point is that governance does not bear a zero-sum relationship to government. Governance cannot be so easily separated from government, insofar as the state is present in each, and, indeed, as states incorporate forms of 'governance' in adjusting to changes associated with 'globalization'. We view 'governance' as an *ideal-typical* concept in two respects: first, it tends to be differentiated from 'government' as if the latter does not itself comprise or constitute relations of rule outside of formal institutional mechanisms, and second, it obscures the role of states in conditioning or constituting governing as part of their authorship of 'globalization'. In the following sections we illustrate these claims through an examination of agricultural regulation.

Governance and agricultural regulation

As a corporate project, globalization projects a vision, rationalized, and institutionalized, as 'governance'. In the agri-food sector we have two recent, complementary, formulations that express the leading role of the USA in envisioning and instituting the global corporate project. In 2001, President Bush proclaimed, on the eve of the WTO Doha Ministerial, 'I want America to feed the world. . . . It starts with having an administration committed to knocking down barriers to trade, and we are.' The following year, the US Secretary of Agriculture, Ann Veneman, envisioned a 'global agriculture [where] future agriculture policies must be market-oriented . . . they must integrate agriculture into the global economy, not insulate us from it' (quoted in IUF 2002: 4). The vision of a *global agriculture* is premised on the superiority of a corporate-dominated world market for foodstuffs over domestic food systems. This premise frames the protocols of governing – in fact, when nations sign on to the WTO Agreement on Agriculture (AoA), they surrender the right to pursue a national food security strategy. In addition, states are forbidden to restrict agri-food imports to protect population or livestock health, without scientific proof from experts recognized by the WTO. Since 'almost all remaining WTO-legal support options require direct payments through the government budget' (Einarsson 2001: 6) rather than public support of domestic production, states with public capacity severely eroded by debt and structural adjustment are at a huge disadvantage. In short, the adoption of the AoA by member states is a concrete instance of the embedding of global governing mechanisms in state policy.

Global mechanisms of governing involve 'market rule', in the sense that states incorporate multilateral (sometimes bilateral) protocols into their policy, acceding to the priority of rules designed ostensibly to open (national) markets. States do not evaporate under market rule. Rather, governing involves considerable formal regulatory intervention in recalibrating policies away from public and towards private goals. There is also a substantive dimension, involving the subjection of producer, consumer and social relations to the price form – where social protections (access to an environmental commons, public goods, subsidies, minimum wages, price controls on staple consumer items, trade tariffs and so on) are regarded as frictions that distort market processes.

We note that market rule does not imply a borderless world market. Rather, it involves the adoption by national governments of policies geared to realizing these substantive dimensions within a particular historic conjuncture. That conjuncture is corporate globalization in a world in which Europe and North America contain the dominant agricultural producing states. Within this conjuncture, the major agricultural producing states of the global South (e.g. Brazil, Argentina, Thailand, and now China), with the potential to dwarf the productive capacity of Europe, North America and Australia, face being locked in to a regulatory system that they did little to help create. Because of the high degree of corporate concentration in all aspects of the agricultural sector, from equipment to seeds to services to intellectual property, in order to develop their 'national agriculture(s)', these countries must rely on capital investment from Northern agribusiness in order to engage in the world market, which is quickly becoming the only viable level of economic participation for producers of bulk commodities. Thus, to the extent that *world* market rule privileges global, over national, economic management, it is based in a highly asymmetrical playing field that privileges Northern agribusiness. We argue that because of the asymmetry, market rule functions to lock in Northern privileges through significant concessions to corporate farming, and to dismantle protections for Southern producers. That is, market rule is not the rule of the market so much as the political construction of markets to serve corporate, rather than public, interests.

The controversy regarding the ascension of market rule is not so much that it seeks to protect economic interests or the rights of private property, but that it acts to canonize these rights and to make them inalienable and unassailable, especially where they jeopardize the public good. This has been called the 'constitutionalization' of market rule – the transformation of neoliberal economic principles from a politically negotiated system of 'treaty law' to a sort of 'higher law' that is 'irreversible, irresistible, and comprehensive' (Howse and Nicolaidis 2003: 74; see also Gill 2000). Howse and Nicolaidis claim that 'Constitutionalism is viewed as the means of placing law, or the rule of law, above politics' (2003: 75). It is clear in the emerging WTO regime that the constitutionalization of neoliberal ideology is an ongoing project. It is realized incrementally through multilateral trade negotiations in the WTO as well as in regional agreements such as NAFTA and the FTAA, but is also contested vehemently in new spaces forged by social movements pursuing alternative visions of global governing. Examining some of the particular agreements forged in the above-mentioned trade negotiations, however, reveals the mechanisms of accomplishment of market rule.

Governing via the WTO

One of the key sites of market rule codification is in the WTO's Agreement on Agriculture (AoA), which completed the Uruguay Round of trade negotiations (1986–1994). The introduction to the agreement, published by the WTO under the heading 'Conceptual Framework', clearly articulates some of the fundamental principles of market rule:

> The Uruguay Round resulted in a key systemic change: the switch from a situation where a myriad of non-tariff measures impeded agricultural trade flows to a regime of bound tariff-only protection plus reduction commitments. The key aspects of this fundamental change have been to

stimulate investment, production, and trade in agriculture by (i) making agricultural market access conditions more transparent, predictable, and competitive, (ii) establishing or strengthening the link between national and international agricultural markets, and thus (iii) relying more prominently on the market for guiding scarce resources into their most productive uses both within the agricultural sector and economy-wide.

(WTO 2000: 5)

This statement reveals several of the ideological underpinnings of market rule. First, increased investment, production and trade are desirable ends in themselves, regardless of their social outcomes. This is indicative of a move away from the development project wherein these goals were justified to the extent that they realized the 'social contract' between governments and citizens. Second, 'national' and 'international' are represented as mutually exclusive categories, contrary to agribusiness' integration of these arenas in its global sourcing operations. Nevertheless, the distinction between 'national' and 'international' serves to encourage domestic policies, such as farm subsidies, that artificially cheapen commodity prices and ultimately serve agribusiness interests. Third, there is the assumption that the market will allocate resources in the most efficient way possible. This is the same logic that privileges a corporate-managed global food security relation over local food sovereignty arrangements on the grounds of global 'comparative advantage', ignoring the grossly inadequate and unequal consequences of a trade-based food security regime (McMichael 2003).

The AoA requires adherence to liberalization. Member states' policies that interfere with trade (including interference with production, price, imports or exports) are subject to reduction and eventual elimination under WTO rules. The agreement extends these principles into the realm of domestic policy via the 'box system' – a mechanism designed to ensure that domestic policies that are not immediately trade-related do not permit states to indirectly circumvent their responsibilities to the regime. In the AoA, all forms of domestic support for producers are placed into one of three 'boxes': amber, blue or green. Measures belong in the amber box if they are judged to be 'trade distorting'. This generally includes policies that affect price, production level or both. The most common examples of amber box policies are payments to farmers that are tied to production, market price supports and product-specific subsidies. A summation of the total amount of support provided by a country via these measures in the 1986 to 1988 period (otherwise known as the 'base period') provides the Aggregate Measure of Support, which is subject to reduction under the WTO. All countries that employ policies such as these have committed to a scheduled reduction specific to that country. There is no negotiating out of reduction commitments for amber box policies.

The blue box contains policies that may be trade-distorting, but are limited in their amount and effect. For example, direct payments under production-limiting programmes make it into the blue box if they are made on a fixed area or yield, or a fixed number of livestock, or if they are made on 85 per cent or less of total production in the base period. By linking these payment requirements to the base period rather than yearly yields, they are considered to be less tied to production than those that are based in yearly production (and relegated to the amber box).[1] In WTO language, 'Production is still required in order to receive these payments, but the actual payments do not relate directly to the current quantity of production' and therefore are judged to be less trade-distorting (WTO 2000: 12).

Finally, the green box contains policies that are judged to be non-trade-distorting. These include conservation measures, disaster relief, agricultural research and extension, and rural development measures, but this most permissive box also includes direct payments to farmers that are not based on any sort of production whatsoever. Many US agricultural support policies have found their way into the green box through creative redirecting. For example, the billions of dollars paid every year by the US government to farmers in decoupled income support payments enjoy residence in the green box. They are not deemed trade-distorting since they are explicitly *decoupled* from production. Producers and/or landowners are paid a set amount by the government for the acreage and production of the eight subsidized commodities[2] during the years 1986 to 1988. A farmer that does not even grow corn will receive a payment every year based on the corn produced on the land she now farms (or even just owns without producing at all) between 1986 and 1988. As strange as this system of farmer supports appears, what the US government has accomplished by placing subsidies in the coveted green box is the decoupling of subsidies from production, an asymmetrical arrangement rewarding corporate farms, and subjecting agricultural output to the price form. The latter constitutes a new 'food regime' of artificially depressed commodity prices geared to reconstructing agriculture globally as a corporate domain (McMichael 2002).

The box system of managing national agricultural policies reveals the principles of market rule and the ways that they are being codified in the emerging and completely new legal system which governs the trade regime. This particular taxonomy defends overtly the ideal of liberalization by eliminating trade-distorting measures while ostensibly preserving the sovereignty of member nations. The agreement states:

> A key objective has been to discipline and reduce domestic support while at the same time leaving great scope for governments to design domestic agricultural policies in the face of, and in response to, the wide variety of the specific circumstances in individual countries and individual agricultural sectors. The approach agreed upon is also aimed at helping ensure that the specific binding commitments in the areas of market access and export competition are not undermined through domestic support measures.
>
> (WTO 2000: 9)

The current structure of the box system in the AoA reveals the true nature of market rule – not as a programme for thorough liberalization, but rather to institutionalize the neoliberal corporate project. For instance, the US$40 billion Farm Bill passed in the US Congress in 2002 is not regarded as an attempt to 'undermine commitments' because the commodity payments that farmers receive are 'decoupled' from production. Therefore, these payments, totalling almost US$33 billion in 2000,[3] are considered to be 'green box' measures, along with the unprecedented commitments to more legitimately 'green' conservation incentives that legitimized the Farm Bill. However, in 2003 the conservation title of the Bill, the Conservation Security Programme (CSP), was essentially eliminated when the US Congress failed to appropriate even a minimal amount of funding for conservation, while maintaining full funding for commodity payments.[4]

It is also significant to note the extent to which the content of the boxes allows much more leeway for the four most powerful agricultural exporting members of the WTO, the US, EU, Canada and Japan (affectionately dubbed 'the Quad') to undermine their commitments than agri-exporting countries of the global South. The Quad has the resources to direct large sums of money into direct support programmes that are deemed non-trade-distorting. Where the US used to rely on production control measures but has since switched to direct income support for farmers, the EU has traditionally relied heavily on price supports that accounted for as much as 80 per cent of the price received by farmers. In 1991, over 90 per cent of the EU Common Agricultural Policy (CAP) funding went towards 'refunds and intervention', or direct price support to farmers. Under WTO rules, governed by the principle of the price form, that is to be reduced to 21 per cent of the CAP, but accompanied by a 60 per cent *increase* in direct decoupled payments to farmers (European Commission, Directorate-General for Agriculture 2001).[5]

Most governments of the global South lack those resources. The only protective measures at their disposal are import tariffs and quotas, which fall under the 'market access' section of the agreement and are the first to go. In addition, *any* non-tariff barrier to market access must be 'tariffied', meaning it must be quantified and then subject to scheduled reductions. Obstacles to market access are considered to be the most egregious examples of government interference in the market, and the principles of market rule require that market logic determine the actions of government, not the other way around. In short, implementing policies that encourage ecologically efficient domestic agriculture (in terms of less resource-intensive productive and distributive capacities), and finance that development through tariffs on foreign imports, is considered worse for the global economy than propping up production of exports made 'artificially' competitive through government income support policies that allow commodities to be sold on the world market well below the cost of production. This is because imposing tariffs on imports results in a net transfer of resources from private to public hands, while subsidizing domestic production results in the reverse.

In other words, global governing, via WTO codifications, is much more effective in *regulating states* than in regulating trade. Of course, this is not to say that the state ceases to be an effective or powerful political actor under market rule. On the contrary, as was clear in the above discussion of the AoA's regulatory 'box system', some states are quite capable of manipulating the system and/or circumventing it by redirecting policies to at least cosmetically meet the letter of WTO regulation, and also by creating 'nonfunded mandates' that comply with WTO rules but do not alter the present domestic subsidy system in any real way, such as the CSP.

Contradictions of global governing

Subsequent to the ratification of the AoA, it quickly became clear that the restrictive market access provisions and the more lenient domestic support requirements heavily weighted the agreement in favour of the Quad powers. Since the WTO Ministerial in Seattle, this arrangement has become unacceptable to many countries from the global South, and it has become the centre-piece of one of many bitter disputes in the WTO, crystallizing as a dispute between the Quad and the 'Group of 20'[6] over reduction commitments written into the AoA (Peine 2003). Essentially, the G-20 refused to lower tariffs if the USA and the EU refused to lower domestic supports. In 2003 the G-20 refused to agree to reduce its market barriers without a restructuring of the current AoA – that includes loopholes large enough for the USA to drive its $40 billion Farm Bill through unscathed. They argue that while the market access provisions of the agreement include a rigorous enforcement schedule, that part of the document pertaining to the reduction of export and domestic subsidies 'fails to set any specific deadlines for doing so or to specify by how much tariffs in Europe, the United States, and Japan will be reduced' (Althaus 2003).

At first glance, the asymmetry of the AoA is puzzling. It is tempting to argue that the reason why domestic subsidies are allowed to slip through the cracks while obstacles to market access are not may be the result of power-mongering by the Quad. There is widespread agreement that the organs of global governing such as the WTO are simply dominated by powerful interests which impose their provisions while the rest of the world unfortunately has to rely on the bureaucratic system, which is predictably slow, cumbersome and costly to navigate (Delich 2002). All of this is undoubtedly true, but there is another reason why the Quad has been so successful in protecting its agricultural supports. Essentially, supports do not threaten the logic of market rule in any profound way. Certainly, they may artificially construct comparative advantage for the highest-cost agricultural producers in the world, but they do succeed in generating the lowest prices for agricultural commodities in history.

Farm prices for the major commodities in world trade have fallen 30 per cent or more since the signing of the Agreement on Agriculture in 1994 (Ritchie 1999), while *The Economist* (17 April 1999: 75) notes that commodity prices are at an all-time low for the last century and a half. Import and/or export restrictions, production controls and state trading, on the other hand, threaten to increase the market price farmers receive for their produce. The core of market rule is laid bare in the fact that this is considered to be 'artificial' price support, while the depression of prices maintained through the Quad's domestic support policies is not. More than simply the subjection of agriculture to the price form, the depression of agricultural prices functions as an indiscriminate weapon of agribusiness against all farmers. Market rule, then, expresses the diametrically opposed outcomes framing the agricultural dispute in the WTO: blatant violation of its espoused principles by Northern agricultural protections, and wholesale expropriation of small farmers in the global South.

Despite the characterization of the G-20's position as progressive and the Quad's as protectionist, both equally serve the larger neoliberal project of which the WTO is currently the most prominent regulatory mechanism. Although the G-20 has demanded greater democracy and transparency in the process along with lowering of agricultural protections in the Quad, the main goal is to facilitate the exploitation of these countries' 'comparative advantage', which typically means low-value, labour-intensive primary commodities, light manufacture or low-skilled services. For Southern beneficiaries of the neoliberal project, a more democratic and transparent WTO is the key to meeting these objectives – which presume a growing labour force of dispossessed farmers. On the other hand, the Quad seeks to maintain its current level of support for its farmers, while forcing down tariffs and import quotas to open up vast potential markets for its artificially cheapened agricultural commodities. And this is what farmer organizations (as opposed to commodity groups such as the ASA) around the world and in the USA see as the real problem: low world prices for agricultural products. Agriculture may be the only sector in which world market prices are generally well below the cost of production. The most extreme example is cotton, with a world market price 57 per cent below the cost of production (Ray *et al.* 2003).

It may seem that these prices, which would be unsustainable without government support, signal an unhealthy and maladjusted market. It is impossible to come to such a conclusion, however, in the face of the explicit preservation of the supports that allow for these prices in the WTO itself. To understand this apparent contradiction it is important to look at who benefits from such low prices. Certainly not the farmers of the Quad, since most farmers would not only prefer earning a living in the marketplace to farming the government, but also are barely staying out of debt even with government loans and income supports. Certainly not governments/ taxpayers, burdened with huge pay-outs. Certainly not consumers, supposedly benefiting from 'cheap food' – in the USA, a box of shredded oat cereal still costs almost four dollars, while the commodity value of the oats in that box is about nine cents.

The fact is that most agricultural products, with the exception of the vegetables sold at farmers' markets, are bought not by the end consumer, but by agribusinesses that store, trade, process and/or package that product. The low commodity price does not necessarily mean cheaper cornflakes, but it does translate into higher profits for General Mills, Kellogg's, ADM or Cargill. Agricultural commodities such as corn, soybeans, tomatoes, feeder calves, cotton, sugar and rice, are 'industrial inputs' for these companies, and so it is in their interest to keep prices for these products as low as possible (Hansen 2003). These firms do not see production control programmes that pay farmers to idle land or store grain in order to prop up prices (similar to the OPEC scenario) as viable policy solutions.

In fact, the 1996 Freedom to Farm Act effectively dismantled land set-aside programmes that had long supported agricultural prices in the USA. Within the next four years, world prices plummeted 40 per cent (Ray *et al.* 2003). It was this collapse in prices and farmer income that triggered ballooning subsidy payments required to keep the entire American farm sector afloat. Because the USA dominates the market for most traded commodities, the low prices of US exported commodities have the power to drive prices down all over the world, which has resulted in the devastation of farmers worldwide. Conservative estimates are that between twenty and thirty million people have lost their land (and relatively secure livelihoods) as a result of trade liberalization (Madeley 2000: 75).

The US accounts for 70 per cent of world corn exports,[7] which shape world prices, and especially those in Mexico, where prices have plummeted 70 per cent, driving waves of farmers off the land (Oxfam 2003: 17). As many as 1.75 million *campesinos* have been dispossessed since the inception of NAFTA in 1994 (Carlsen 2003). At the same time, government deregulation has caused a threefold increase in the real price of corn tortillas, resulting in widespread food crises in both rural and urban areas (Bensinger 2003; Oxfam 2003: 19). This pattern has been repeated in countries all over the world which lack the resources to provide the kinds of income support that keep US farmers on the land. Even there, 33,000 US farmers, with less than US$100,000 annual income, have left the land since 1994 (*Public Citizen* 2001: ii–iv). Under this regime, those farmers in the world that are most efficient are of no particular interest to agribusiness, so long as commodity prices remain low. Liberalization, theoretically bringing markets to equilibrium, is clearly not the intended goal, even though it provides the rhetoric of market rule.

Corporate governing

Through examination of the AoA, it becomes clear that market rule is less about freeing trade or markets than it is about consolidating the power of agribusiness to organize its money and commodity circuits. Certainly WTO rhetoric is that the generalization of agri-exporting would render agricultural protections unnecessary, because of expanding global food flows. *Public Citizen* observed, with regard to NAFTA and WTO policies: 'Proponents of the legislation contended it would make farming more efficient and responsive to market forces; in reality it essentially handed the production of food to agribusiness' (2001: 16). In Europe, the bulk of emergency taxpayer assistance went to the largest farms. As governments that could afford farm support ran up huge relief bills, agribusinesses took the opportunity to restructure, with input and output industries consolidating:

> within and across their narrow sectors and (creating) alliances with other food industries to encircle farmers and consumers in a web ... from selling seeds and bioengineering animal varieties to producing the pesticides, fertilizers, veterinary pharmaceuticals and feed to grow them to transporting, slaughtering, processing and packaging the final 'product'.
> (*Public Citizen* 2001: 19)

Such 'food chain clustering' (Heffernan *et al.* 1999) exemplifies general *global* restructuring trends, where, for example, 60 per cent of foreign direct investment in 1998 involved cross-border mergers and acquisitions (*Public Citizen* 2001: 19). The Canadian National Farmers' Union testified in 2000 that 'almost every link in the chain, nearly every sector, is dominated by between two and ten multibillion-dollar multinational corporations' (quoted in *Public Citizen* 2001: 20).

Through the device of cross-border operations, global firms exploit food market asymmetries between North and South, undercutting Northern entitlement structures and their institutional supports by optimizing the strategy of global sourcing. Cargill's Director of Public Policy, Bryan Edwardson, claims: 'Whether it's Brazilian or American soybeans, the world needs both' (Diaz 2003). The apparent equanimity of this statement is mediated somewhat by the more direct statement by Warren Staley, Cargill's CEO, who notes that '[Brazil] has very free markets. They have internal infrastructure challenges that can only improve. So Cargill makes a choice. Are we going to put a whole lot more money and grain and agriculture processing in the US, or in Brazil...?' (Daniel 2004). In other words, the world needs American soybeans for now, but the 'choice' Cargill has to make has the power to render the American soybean obsolete.

Ultimately, global sourcing depends on the political restructuring of the market. Market rule is predicated on protecting corporate rights first, where liberalization may be a condition, a consequence or a non-starter, depending on the circumstances. In some instances, corporate protection means instituting more rules, not fewer, and depends on enforcement by the state. These include, in particular, rules that deepen corporate power via market mechanisms. Two exemplary cases are, first, the notorious Chapter 11 of

NAFTA, which allows foreign corporate compensation for profits forgone (e.g. if a state forbids privatization of a service, or decommodifies a natural resource); and second, intellectual property rights (via the TRIPs protocol in the WTO), where corporations expect the state to *create* a market where one did not exist before. The 'market' for intellectual property, essentially, can never be deregulated because it is the law itself that gives rise to this unique form of property.

Conclusion

Market rule, then, depends on state intervention in some sectors just as desperately as it strives to expunge it from others. Despite arguments to the contrary, the state remains critical to market rule.[8] This is clear from our analysis of the asymmetries of the AoA. Not only does the asymmetry in the implementation of WTO protocols discussed above promote the price-lowering regime that serves the interests of agribusiness, but it also maintains the government supports that sustain one of the most socially consequential economic sectors in four of the most politically powerful states. Most of the attention given to the conflict between the G-20 and the Quad casts it as a political dispute framed in Third World/First World terms. However, to the extent that G-20 states demand fulfilment of liberalization promises in greater access to Quad agricultural markets, they are essentially playing the same game – promoting the agri-export regime. And this is a game that only agribusiness can play, since its 'comparative advantage' is its monopoly on global sourcing and production inputs and outlets.

Many governments in the global South face an impasse, where foreign currency requirements to service their debt intensify agri-exporting, or debt-rescheduling involves 'second generation structural adjustment' (linking credit to 'good governance' such as privatization), generate legitimacy crises as they withdraw land from staple foods, or public services from citizens. The growing legitimacy crisis expressed itself in the rebellion at Cancun against the implications of corporate governing. The organization of the G-20 was the palpable result, but its potency is likely to be undermined by the introduction of *bilateral* manoeuvres by the Quad powers to tempt individual Southern states into the Faustian bargain of access to Northern markets.

Market rule, then, is not about the exile of politics from the economy. Insofar as the state is implicated in market rule, its dual role as protector of the national interest and facilitator of trade liberalization is potentially in conflict, unless the national interest is identified with trade liberalization. In the global North, this expresses itself in mapping the public's interests in a safe, plentiful, cheap and diverse food supply on to the interests of corporations in maximizing profits. A subtext is often the fruits of empire, or at least sustained access to offshore sources of food (cf. Mintz 1986). In the case of agriculture, it is argued that liberalization (of trade, as well as capital mobility) will lead to greater productivity and therefore cheaper food for the global population. As we have suggested, neither outcome is certain in a corporate global agriculture, based in unsustainable monocultures and managed by 'consumer states' with the power to orchestrate market rule through institutions of governing such as the WTO.

In conclusion, we note that these trends are increasingly contested by the proliferation of alternatives such as fair trade schemes, community supported agricultures, land repossession movements such as the Brazilian landless workers' movement, and the coordination of 'food sovereignty' demands through the recently formed transnational farmers' movement, the *Via Campesina* (Desmarais 2002). Such resistances express the social and ecological contradictions of market rule, and expose the rhetoric of governing that seeks to legitimize a state-authored global project of corporate agriculture. These movements also challenge the 'neutrality' of the market, piercing the veneer of inevitability that sheaths the neo-constitutional project. The neoliberal rhetoric of 'global governing' proclaims a historically self-evident world trajectory, with the market as organizing principle. In such a neo-Darwinian world, the liberal economic model has triumphed over central planning, socialism and other attempts at political regulation of the economy, and represents alternatives as historical anachronisms. The extent to which this argument may ring true, however, is not at the centre of the governance debate. The real question is *where, when and in whose interests* government regulation will continue to function in managing the global economy. This is the real subject of trade negotiations, and the emerging consensus promises great consequences for even the farthest removed from the hotel conference rooms of WTO ministerials.

Notes

1 The definition of the base period, however, has raised a great deal of controversy due to the fact that government subsidies as well as production levels were at all-time highs during these years, and many governments automatically conformed to their reduction schedules without actually having to reduce support at all.
2 Corn, wheat, soybeans, barley, sorghum, oats, rice and peanuts.
3 USDA Economic Research Service, available at http://www.ers.usda.gov/ publications/agoutlook/aotables/feb2004/aotab30.xls. Total green box payments for 1999 totalled almost $50 billion (WTO, available at http://www.wto.org/ english/tratop_e/agric_e/negs_bkgrnd19_data_e.htm).
4 National Campaign for Sustainable Agriculture, available at http://www.sustainableagriculture.net/NCSACSPAction.php.
5 Interestingly, the report focuses much more heavily on efforts by the European Commission to *cut prices*, not to reduce government support for the agricultural sector.
6 The G-20 (though sometimes called the G-X since no one seems to know any longer exactly how many countries count themselves as members) is a group of countries from the global South formed to leverage power away from 'the Quad'. The G-20 is led by Brazil and includes India, China, South Africa, Egypt and Argentina.

7 And 70 per cent of those exports are accounted for by just two companies: Cargill and Archer Daniels Midland (ADM).
8 It is important to note that states are not all equal, and that the WTO regime suffers from sometimes quite profound differences among states regarding social priorities and institutional capacities. For analysis of the challenges faced by poorer countries built into the Dispute Settlement Mechanism of the WTO, see Delich (2002). For a cogent analysis of the obstacles posed by distinctive regulatory frameworks to the WTO regime with respect to multilateral agreements regarding genetically engineered organisms (GEOs) between the US and the EU, see Buttel (2003).

References

Althaus, D. (2003) 'WTO subsidies draft angers poorer nations', *The Houston Chronicle*, 13 September.

Arrighi, G. (1982) 'A crisis of hegemony', in S. Amin, G. Arrighi, A.G. Frank and I. Wallerstein (eds) *Dynamics of Global Crisis*, New York: Monthly Review Press.

Bensinger, K. (2003) 'Mexican corn comes a cropper', *The Washington Times*, 9 September.

Buttel, F. (2003) 'The global politics of GEOs. The achilles' heel of the globalization regime?', in R.S. Schurmann and D.D.T. Kelso (eds) *Engineering Trouble. Biotechnology and Its Discontents*, Berkeley: University of California Press.

Carlsen, L. (2003) 'The Mexican farmers' movement: exposing the myths of free trade', *Americas Program Policy Report*, Silver City (NM): Interhemispheric Resource Center.

Daniel, C. (2004) 'The Cargill approach: We don't lobby. We go and share information', *The Financial Times*, 26 February.

Delich, V. (2002) 'Developing countries and the WTO dispute settlement mechanism', in B. Hoekman, A. Mattoo and P. English (eds) *Development, Trade, and the WTO: A Handbook*, Washington, DC: IBRD/World Bank.

de Senarclens, P. (1998) 'Governance and the crisis in the international mechanisms of regulation', *International Social Science Journal*, 155: 91–104.

Desmarais, A.A. (2002) 'The Via Campesina: consolidating an international peasant and farm movement', *Journal of Peasant Studies*, 29, 2: 91–124.

Diaz, K. (2003) 'Brazil: The new breadbasket', *Star Tribune*, 7 March, available at http://www.startribune.com/stories/484/4647358.html.

Einarsson, P. (2001) 'The disagreement on agriculture', available at www.grain.org/publications/mar012-en.cfm.

European Commission, Directorate-General for Agriculture (2001) *EU Agriculture and the WTO: Towards a New Round of Trade Negotiations*, Brussels: European Commission, Directorate-General.

Gill, S. (2000) 'The constitution of global capitalism', Paper prepared for the International Studies Association 41st Annual Convention, Los Angeles, CA, 14–18 March.

Hansen, J.K. (2003) 'Trade debate over ag subsidies misguided says US trade advisor', Press release from the Nebraska Farmers' Union, available at http://www.nebraskafarmersunion.org/.

Heffernan, B. *et al.* (1999) 'Consolidation in the food and agriculture system', Report to the National Farmers' Union, available at www.nfu.org/whstudy.html.

Howse, R. and Nicolaidis, K. (2003) 'Enhancing WTO legitimacy: constitutional-ization or global subsidiarity?', *Governance: An International Journal of Policy, Administration, and Institutions*, 16, 1: 73–94.

IUF (2002) 'The WTO and the world food system: a trade union approach', Geneva: International Union of Food, Agricultural, Hotel, Restaurant, Catering, Tobacco and Allied Workers' Associations.

Jessop, R. (1998) 'The rise of governance and the risks of failure: the case of eco-nomic development', *International Social Science Journal*, 155: 29–46.

Kazancigil, A. (1998) 'Governance and science: market-like modes of managing society and producing knowledge', *International Social Science Journal*, 155: 69–80.

Kooiman, J. and Van Vliet, M. (1993) 'Governance and public management', in K. Eliassen and J. Kooiman (eds) *Managing Public Organizations*, 2nd edn, London: Sage.

McMichael, P. (2002) 'La restructuration globale des systems agro-alimentaires', *Mondes en Developpment*, 30, 117: 45–54.

McMichael, P. (2003) 'Food security and social reproduction: issues and contradic-tions', in I. Bakker and S. Gill (eds) *Power, Production and Social Reproduction*, New York: Palgrave Macmillan.

Madeley, J. (2000) *Hungry for Trade*, London and New York: Zed Books.

Mintz, S. (1986) *Sweetness and Power. The Place of Sugar in the Modern World*, New York: Vintage.

Oxfam (2003) 'Dumping without borders: how US agricultural policies are destroy-ing the livelihoods of Mexican corn farmers', *Oxfam Briefing Paper,* 50, available at www.oxfam.org.

Peine, E. (2003) 'Seeds of dissent', *The Bookpress*, November: 3.

Prakash, A. and Hart, J. (eds) (2000) *Globalization and Governance*, London and New York: Routledge.

Public Citizen (2001) 'Down on the farm: NAFTA's seven-year war on farmers and ranchers in the U.S., Canada and Mexico', *Public Citizen*, 26 June, Washington, DC, available at www.citizen.org/documents/ACFF2.pdf.

Ray, D., De La Torre Urgarte, D.G. and Tiller, K.J. (2003) *Rethinking US Agricul-tural Policy: Changing Course to Secure Farmer Livelihoods Worldwide*, Knoxville, TN: Agricultural Policy Analysis Center of the University of Tennessee.

Ritchie, M. (1999) 'The World Trade Organization and the human right to food security', Presentation to the International Cooperative Agriculture Organization General Assembly, Quebec City, 29 August.

Ruggie, J.G. (1998) 'Epistemology, ontology, and the study of international regimes', in J.G. Ruggie (ed.) *Constructing the World Polity*, London and New York: Routledge.

Stoker, G. (1998) 'Governance as theory: five propositions', *International Social Science Journal*, 155: 17–28.

World Trade Organization (2000) *Agriculture*, Geneva: WTO.

3 Re-shaping the state

Global/local networks of association and the governing of agricultural production

Lynda Cheshire and Geoffrey Lawrence

Introduction

In recent decades, the governing of agricultural production has been profoundly transformed. Where, in the past, primary producers and their associated rural communities were protected from the vagaries of the world market through state-based regulatory frameworks of trade tariffs, pricing subsidies and quota protection, now these frameworks are being progressively dismantled, forcing farmers to compete on a global scale for increasingly fickle markets. In spite of these shifts being executed under the banner of free trade (McMichael and Lawrence 2001: 154; see also Peine and McMichael, Chapter 2, this volume), and, with farmers being promised increased freedom in the marketplace, the reality for many producers has been their subjection to new demands from transnational corporations, supranational organizations such as the European Union, environmental bodies, and a host of groups which contest claims that rural spaces are exclusively sites of production. Unable, or unwilling, to regulate the interactions of these various stakeholders, the nation state has come under much scrutiny as questions are raised about its continued relevance in an era where non-state actors at the global, national and local levels are increasingly influential in shaping the activities of agri-food producers. While few would argue that nation states have been rendered meaningless or powerless by these changes, it is worth considering how the emergence of these new actors into the agri-political arena have changed our understanding of the modern state and, by implication, the way in which agricultural production is governed in contemporary societies.

The manner in which the activity of governing is 'escaping the categories of the nation state' (Held 1991: 204), shifting out of the hands of a sovereign authority and into various arenas comprising state and non-state actors, is conceptualized increasingly by a number of scholars in terms of a shift from government to governance (Stoker 1998: 17). The factors underpinning these changes are varied, but two related processes over the last half-century – globalization and the rise of neoliberalism as a guiding rationality for government action – have been particularly significant. On the one hand, the establishment of multilateral trade agreements and other supranational regulatory regimes have transferred power 'upwards' into the hands of bodies such as the World Trade Organization (WTO) and the International Monetary Fund (IMF). On the other hand, the perceived failure of the welfare state to cope with the burgeoning needs of the population for a growing range of services, coupled with the increasing demands of citizens for greater input into the policy-making process, has shifted power 'downwards' to citizens and communities. While it is widely accepted that the 'rescaling' of state power has created a multifarious landscape of institutional forms at multiple scales (Jones and MacLeod 1999: 299), the complexity of these new arrangements cannot be captured fully by concerns with vertical shifts in power alone. Not only do such analyses overlook the blurring of horizontal distinctions between state and civil society that is also occurring under contemporary arrangements of governing; they also tend to prioritize linear conceptions of power as it is exercised from the top down. The outcome in both cases is a silence on how agri-food producers, either collectively or individually, form networks of association to reshape and transform these power relations according to their own objectives (Herbert-Cheshire 2003).

The purpose of this chapter is fourfold. First, it examines the dual processes of globalization and state restructuring that have fostered the emergence of a new politics of regulation in the agri-food sector, and shows how a range of non-state actors have become significant to our understanding of contemporary governing arrangements. Second, we consider the implications of these new arrangements for the on-farm activities of primary producers in developed nations such as Australia. Third, it is argued that as power is reconfigured, both vertically *and* horizontally 'beyond the state' (Rose and Miller 1992), traditional conceptions of the state as *the* site of political power are no longer useful for making sense of the 'fragmented maze' (Stoker 1998: 19) of new forms of power relations. Instead, this chapter draws upon a growing body of work, inspired partly by a Foucauldian 'analytics of power' (Foucault 1978: 82) and partly by earlier versions of Actor-Network Theory (ANT) (Callon and Latour 1981; Latour 1986; Law 1992), to consider power as the outcome of collective actions exercised through networks of associations. The appeal of this approach is that it provides the conceptual tools for deconstructing 'powerful' actors such as the state so they may be viewed as contingent networks of disparate actors and agencies. It also offers a way of understanding how micro actors – rural citizens, farmers and local action groups – interact with these networks to engage with, and potentially reshape, the practices and outcomes of rule. Fourth and finally, this chapter draws upon two brief examples of how farmer agency is exercised to consider how these networks of association have the potential to challenge the current neoliberal globalization path.

Globalization and the rise of neoliberalism: new forms of rule in agricultural production

With economic processes becoming increasingly transcontinental, with instantaneous communication, and with growing recognition among the world's citizens that they physically share the finite entity of the globe, Robertson's (1992: 8) conceptualization of globalization as both 'the compression of the world, and the intensification of consciousness of the world as a whole' seems both apt and compelling. In the agri-food sector, the impacts of globalization – along with the introduction of neoliberalism in contemporary government thinking – are clearly apparent. Together, these processes have fundamentally restructured the way in which farming is practised and governed.

The Keynesian welfare state which survived until the 1960s was one that had significant autonomy via its control of capital, manipulation of macro-economic policies and national regulation. In countries such as Australia, these state-enacted measures were designed to expand exports, increase the productivity of farmers, stabilize farm incomes, and achieve economic and social equity between rural and urban dwellers. In an effort to achieve this, a suite of policy settings, including import restrictions, output subsidies and statutory marketing, were established (see discussion in Lawrence 1987). This represented a very strong state-directed attempt to influence the market conditions under which primary producers operated.

Accelerated economic globalization occurred through the 1960s and 1970s with US corporations creating foreign subsidiaries via take-overs and mergers of firms in the manufacturing industries (Green and Wilson 2001). From the 1980s, the politics of protection were quickly replaced with the politics of deregulation (Green and Wilson 2001; Lawrence 1987) that helped to facilitate global trade and to secure a central place for trans-national corporations within it. In Australian agriculture, the earlier welfare state model of a tariff-protected and subsidized farmer delivering commodities to a statutory board, and having that organization sell the product under monopoly conditions on the domestic and international markets, gave way to free-market conditions. Under these new conditions, producers would link directly with corporate firms and, for some industries such as chicken and vegetable production, under contract arrangements (Burch *et al.* 1996). These moves extended the already significant influence of corporate capital in the upstream (input supply) and downstream (processing, storing, packing and selling) components of agri-food activity (Heffernan 1999). In the agro-chemical field, for example, five companies worldwide account for some 60 per cent of the global pesticide market. Similarly, among Australia's broadacre farmers in 1997, some 30 per cent of farmers were producing 70 per cent of output (Robertson 1997). This concentration and centralization of capital continues apace, with the industrialization of agriculture being driven by corporate capital under a mantle of neoliberal policies derived from the WTO and readily embraced by nation states (Heffernan 1999; McMichael 2003; Teubal and Rodriquez 2003).

For many local producers this high level of corporate concentration can lead to subsumption: a dependence upon off-farm entities for what are increasingly expensive inputs to agriculture – one of the most important of which is capital (see Mabbett and Carter 1999). Farmers who borrow capital are often required to abide by the finance lender's conditions, such as the restructuring of on-farm activities, the shedding of labour (normally through the purchase of agribusiness-produced machinery), and specialization in a certain commodity. Such actions tend to reinforce the productivist system of farming, the two major implications of which have been rural community decline and environmental degradation (Gray and Lawrence 2001).

While the power of global capital to determine the conditions of production is clearly enhanced under these new arrangements – largely at the expense of small-scale producers – a second, related, process of rescaling has also taken place which emphasizes the apparent freedom of those same producers to set the terms of their participation in the new economy (McKenna *et al.* 1999). With contemporary Western governments no longer willing, or able, to provide an unconditional safety net of state protection, farmers and other rural dwellers have been reconstituted as 'active citizens' and judged capable of taking responsibility for a range of activities that were previously considered to lie within the province of the state. Such moves are frequently associated with attempts to 'roll back' the state; yet, as Woods (2004) argues, this transfer of responsibility to citizens is also due to citizens' perceptions that the state has failed to deliver the outcomes they desire, and that it is now up to them to find their own solutions. The convergence of these processes in governmental policy has led to the promulgation of new forms of governing that rely upon discourses of self-help (Herbert-Cheshire 2000) or self-reliance (Higgins 2002). In practice, this has resulted in a proliferation of community, and other non-governmental, organizations that work 'in partnership' with the state, as well as a broad range of government programmes that espouse the virtues of individual autonomy, personal responsibility and self-governance.

Government thinking about agricultural production and resource management clearly reflects this trend. In agricultural policy, for example, the Rural Adjustment Scheme that emerged in the 1970s to encourage amalgamation of properties (and so increase returns on scale in farming, and remove the least viable producers) has been viewed by Higgins (2002) as an important refocusing of how, and by whom, Australian agriculture is governed. Viable producers are regarded as those who can embrace good financial management without the need to rely upon the state to guarantee them a future in farming. Such producers should, for example, calculate that drought is a normal occurrence which must be addressed at the property level – that is, within a strategy of risk management – rather than relying upon a benevolent state for hand-outs in hard times (Higgins 2002). In a similar vein, the National Landcare Programme may be viewed as a state-sponsored attempt to counteract some of the negative environmental consequences of globalized agriculture by encouraging producers to work in partnership with the state (see Lockie 1999). While 'empowerment' and 'community ownership' have been central motifs of the Landcare Programme (Martin 1997: 45), Lockie and others (see e.g. Martin 1997) have suggested that such schemes represent a mechanism by which state power may be employed in a direct yet subtle manner to shape the actions of producers so that they conform to a neoliberal, productivist agenda.

Rethinking the state: networks of association

It is readily apparent from the above discussion that the dual processes of globalization and the demise of welfare state-ism have led to a profound transformation in the institutional landscape of agricultural regulation. With the emergence of a global economy – one dominated heavily by transnational corporations (TNCs) – power has shifted upwards to supranational bodies that place limits on the ability of any single state to intervene in the workings of the global system. At the same time, a process of restructuring within the nation state has focused renewed attention upon the obligations of citizens to improve their own conditions of existence. As a result, therefore, contemporary forms of rule are said to have been 'rescaled', in an upward and downward direction via the establishment of new, vertical structures of governing (Brenner 1998; Edwards *et al.* 2001). Precisely what role the nation state may play in these new structures is a matter of much debate. Yet most authors have generally agreed that while the sovereignty of the nation state has been challenged in recent years (Camilleri and Falk 1992), the nation state itself continues to be significant, having driven much of this reterritorialization process in its support of the neoliberal, globalization trajectory (Almås and Lawrence 2003).

Theories of state rescaling are useful to the extent that they capture the changing geography of contemporary governing. However, their tendency to focus on linear processes of vertical restructuring means that the changing relationship between the state and the individual remains largely unexplored. Under contemporary arrangements of governing, power has not merely been rescaled but reconfigured across horizontal distinctions between state and civil society (Counsell and Haughton 2003: 226) as a growing range of actors and agencies, traditionally considered to lie outside the domain of the 'political', come to play an increasingly important role in the exercise of political power. The growing emphasis on state–community partnerships, and the inclusion of local people in the formulation and delivery of government policies through programmes such as Landcare, illustrate the extent to which the activity of governing has moved beyond the realm of the state. Such shifts have prompted writers such as Rose (1993, 1996) and others (Barry *et al.* 1996) to describe the state as having been 'degovernmentalized' as it increasingly devolves responsibility (and sometimes decision-making power) to these other agencies. Rather than assembling all their regulatory mechanisms into a single entity known as the state, therefore, political authorities are exploring new ways of governing that rely upon the skills and expertise of a range of non-political individuals, agencies and techniques – such as self-help consultants, community development officers, capacity building experts and financial counsellors – who are enrolled into a complex network of power relations. It is through this network of state and non-state actors that, on the one hand, the actions of individuals at the micro-level are connected to the broader sociopolitical objectives of governmental authorities (Rose 1993: 286) and, on the other, that an appearance of the state as a single, powerful actor is maintained (Rose and Miller 1992: 176).

From this perspective, the apparent power of the state is the result of various disparate but interdependent activities and agencies, which span different levels of government and different fields of society, and which have been assembled loosely into what may be described as a 'functioning network' (see Miller and Rose 1990: 4). The theoretical influences behind this approach are readily apparent and combine elements of a Foucauldian 'analytics of power' (Foucault 1978: 82) with the early writings of ANT (see e.g. Callon and Latour 1981; Latour 1986; Law 1992). According to the former, power is the 'resultant', not the cause, of action (Rose 1996: 43); for the latter, it is the outcome of a series of *collective* actions by a range of actors enrolled in an actor network. Brought together, these analytical approaches show that agency is not an inherent property of an individual, but becomes effective when people, objects or similar entities are enrolled into an ordered network according to a particular scheme (Latour 1986: 284). What this means is that seemingly powerful actors, such as the state, cannot be seen as a single site of political power, but rather as a configuration of power relations between independent actors who participate in the activity of government for their own ends. Once 'deconstructed' in this way, the power of the state no longer seems to be so monolithic, but much more contingent upon the stability of the networks and the kinds of interaction that take place between its constituent actors (Murdoch 2000). This is because the process of enrolment is never absolute but is subject to a process of 'translation' (Callon 1986: 223) as each actor in the network seeks to redefine the interests of others according to his or her own objectives. The result is a continuous transformation of policies, goals or outcomes as they are encountered by a range of actors who 'slowly turn [them] . . . into something completely different as they [seek] . . . to achieve their own goals' (Latour 1986: 268). Hence the act of governing becomes 'precarious' (Clark and Murdoch 1997: 45) and more likely to be subject to change and transformation.

The idea that power is exercised through networks of association has become increasingly popular in recent years as a way of capturing the complexity of contemporary power arrangements. Although Woods (1997) points out that ANT fails to explain why some actors or entities are enrolled into an actor network at the expense of others (thereby implying that power must be pre-existing), the framework still has some appeal, particularly by virtue of the insights it provides into the way coalitions are constructed and dissolved. Importantly, these coalitions do not operate at a single spatial scale, but span different tiers and domains at the global, national and sub-national levels. Actors from these various domains are linked through these networks into what Murdoch and Marsden term 'actor-spaces' (1995: 369): arenas comprising both local and non-local, political and non-political actors, resources and representations that bind micro issues to national and international events. To a large extent, this network analysis has been favoured by governmentality-inspired theorists who have shown how the conduct of local actors is governed according to the imperatives of late capitalism via a seemingly non-political network of actors, techniques and programmes (see e.g. Herbert-Cheshire 2000; Higgins 2002; Lockie 1999). They have argued that, through this network, political authorities are able to exert considerable influence over local events, decisions and actions 'at a distance' (Latour 1987: 219).

While an analytics of power makes room for the possibility of a 'reverse relationship' (Cooper 1994: 437) in which power may also be extended by those who are subject to it (see also Dean 1994), few contemporary theorists have developed this argument to any great extent.[1] In this sense, the idea that power might be exercised 'from below' is frequently overlooked, in spite of a few broad statements by Rose and Miller (1992: 190) that 'each actor, each locale, is the point of intersection between forces, and hence a point of potential resistance'. How this resistance may occur, what forms it takes, what its effects are, and how other actors may respond to it are all issues that are rarely addressed. Paradoxically, there has also been a tendency within the current agri-food literature to deconstruct the power of the state in the manner described above, while still prioritizing state power in its analysis of advanced liberal forms of rule. Indeed, Lockie and Kitto (2000: 12) have pointed out that the same main actors and institutions are frequently privileged in contemporary accounts of agricultural governance: usually the state as it governs on-farm practices according to the sociopolitical ambitions of late capitalism, or the increasingly powerful food and agrochemical companies that subject producers to the imperatives of global profiteering.[2] Our discussion at the beginning of this chapter indicates the significant, and often detrimental, impacts these relations of rule have upon the economic and social well-being of primary producers. Yet this is not to suggest that producers cannot establish their own networks of association to reshape and transform elements of these relations according to their own objectives (Halpin 2003). Nor is it necessarily the case that the enrolment of local producers in the networks of government (via state–community partnership programmes, for example) prevents them from attempting to redefine the interests of that network, thereby causing those interests to change continuously.

Reshaping the state: two examples

In the remainder of this chapter we consider two very brief examples from the agri-food literature that examine the various ways in which global strategies of power may be adapted and affected by instances of local agency. The significance of these, and other, accounts (see e.g. Callon 1986; Clark and Murdoch 1997; Cox 1998; Murdoch and Marsden 1995) is that by rejecting any analytical distinction between macro and micro actors, they give the scope to explore the exercise of power by the rural citizen, the primary producer and the local action group as they do by a 'macro actor' such as the state. In doing so, they also adopt an actor network analysis to show how local people seek to build their own networks for 'acting at a distance' and contest or transform state policies. In the first case study, this occurs through the establishment of associations between local producers and various other actors, resources and entities which are translated into a common, and powerful, set of interests. The second case study illustrates how key government programmes are 'translated' as networks are constructed between state agencies and local producers that enable state experts to be counter-enrolled into local, rather than state, agendas. The capacity of these short actor networks to influence decision-making may be less than those of longer networks with ostensibly more actors and resources at their disposal. Nevertheless, as the empirical examples reveal, their activities are notable by virtue of the challenges they pose to the political rationalities of the state and the opportunities that arise, however small, for local ambitions to be articulated and realized. Both examples also show how 'connected chains of association' (Donaldson *et al.* 2002: 211) are established between actors at various spatial scales to link localized acts of resistance with national and global politics.

The first example comes from New Zealand where McKenna (2000) examines the contestation by local pipfruit (primarily apple and pear) growers of the logic of the globalization process, and their subsequent resistance to the proposed deregulation of the New Zealand apple industry. The ferocity with which New Zealand pursued an agenda of neoliberal reform from the mid-1980s onwards, resulting in market liberalization, industry deregulation and free trade, has been well documented (see e.g. Kelsey 1995). In light of this reform, it is somewhat paradoxical that New Zealand's apple industry was the last remaining industry with a state-protected, single desk export cooperative – that of the New Zealand Apple and Pear Marketing Board (NZAPMB) (McKenna 2000: 369). In its 1998 budget, however, the New Zealand government announced that all remaining producer boards, including the NZAPMB, would be deregulated, thereby exposing New Zealand pipfruit growers to an unregulated and relatively volatile global market. As well as potentially undermining their financial viability, pipfruit growers were concerned about the flow-on effects of deregulation to associated rural communities, and the population decline, rural service withdrawal and general rural downturn that might ensue.

While McKenna does not draw explicitly on an actor-network analysis to make sense of the way local pipfruit growers responded to the proposed deregulation of their industry, the formation of the anti-deregulation lobby group, United Fruit, and its attempt to link rural industry restructuring to a broader critique of the neoliberal restructuring process, indicates the composite power of the networks that were subsequently established. Indeed, McKenna herself argues, in a manner similar to Murdoch and Marsden in an earlier ANT-inspired paper (1995), that the networks of association established between local growers and other, regional and national, actors connected local action with wider power networks and 'action at a distance' (McKenna 2000: 377). Initially established in 1997 to promote the welfare of pipfruit growers in the Nelson region of New Zealand, United Fruit broadened its remit after the deregulation announcement, eventually becoming 'one of the most vociferous and organized public lobby groups on rural and deregulation issues' (McKenna 2000: 371). Rather than simply emphasizing the localized impacts of deregulation upon Nelson growers, United Fruit linked the issue of industry restructuring to, and used it as an example of, a much more general critique of the logic of the globalization process, and the ideologically driven nature of the New Zealand government's adjustment experiment. In doing so, it enrolled the support of a broader audience, which not only shared its concerns over the future of rural industries and communities, but also opposed neoliberal economic reform in general. As well as apple growers from Nelson and other fruit-growing regions, primary producers from a range of other industries, as well as rural community representatives and urban voters making a stance against central government policies, were brought together in a loosely bound actor network to contest the process of structural adjustment. Ostensibly, the association of these various actors into an actor network around the issue of deregulation enhanced the power capacity of United Fruit to pressure the state into rethinking its policy focus. To some extent this was successful, since the combined efforts of United Fruit's media campaigns and public demonstrations prompted the Minister for Agriculture to back down from the deregulation agenda only two months after it was announced (McKenna 2000: 376). As McKenna acknowledges, however, this apparent 'win' over the discourse of neoliberalism was not uncontested and neither can it be attributed directly to the actions of United Fruit. Nevertheless, what can be learned from this example is that actors other than state agencies may also form networks of association to bring new actor networks into being that have the potential to engage with, displace and possibly transform those of the state.

The second example is provided by Herbert-Cheshire (2003) and seeks to show that although agricultural governance is frequently directed by the state in an advanced liberal manner, the potential still remains for governmental strategies of power to be translated and transformed by local forms of action (see Herbert-Cheshire (2003) for a more detailed analysis of this case study from an actor-network perspective). Based upon an analysis of the interactions between a local farming organization and state agencies in rural Queensland, this case study highlights more clearly how the enrolment of other actors increases an individual's capacity to act, and how 'connected chains of association' (Donaldson *et al.* 2002: 211) link localized acts of resistance with the governmentalities of the state (Herbert-Cheshire 2003: 465). The case study site in question is one, like many towns and regions in rural Australia, where poor commodity prices, a lack of adequate rainfall, declining services and high debt levels among primary producers have created a situation of economic, environmental and social uncertainty. It is under these circumstances that, in 1995, a group of local farmers established the East Warmington Revival Committee. Believing that prosperity would flow to the region if only resident farmers could become more viable, the group set about pursuing alternative and more profitable rural industries – the most promising of which was a proposed linseed and flax industry. In spite of the group receiving a high level of in-kind support from the local office of the Queensland Department of Primary Industries, the project was hindered by a lack of funding. When approached by the Commonwealth Department of Transport and Regional Services, therefore, to pilot its newly released funding programme – the Rural Plan – the group agreed. The fact that the money was limited to strategic planning and capacity-building activities and, hence, represented a significant departure from the group's original ambitions was compensated by the AUS\$100,000 sum that accompanied this agreement.

However, some members of the group were unhappy with the change of focus, preferring to channel their energies into more practical activities, such as getting the linseed/flax project up and running. While they initially manifested their disapproval by engaging in arguments with other members, or avoiding group meetings, one detractor shifted his position from impeding the planning process to displacing certain elements that did not suit his needs. The opportunity for this to occur arose when a representative from the Federal government travelled to East Warmington to assist the Revival Committee with writing the funding application for the Rural Plan (see Herbert-Cheshire 2003). Recognizing that this was a chance to exert influence over the planning process and gain state support for his linseed/flax project, the farmer sought to counter-enrol the departmental representative and persuade him that the project was worthy of Federal government support. The incorporation of a Federal actor into the local network of the farmer was successful at two levels. First, the farmer was able to reshape the planning process according to his own objectives and to secure AUS\$16,000 from the grant for the linseed/flax project. Had the Canberra representative not made the journey to East Warmington, the opportunity for him to be enrolled into the farmer's scheme would not have been possible and the powerful set of associations between the state's objectives (to pilot a 'successful' funding scheme) and those of the East Warmington group (to establish a viable rural industry) would not have been established. Indeed, the original purpose of the Canberra representative's visit had been to promote the Rural Plan to the East Warmington Committee, yet the enrolment of the dissenting farmer into these activities enabled the farmer to manipulate the process for his own benefit. At a broader level, moreover, it is significant that in 2000, the Department of Transport and Regional Services replaced the Rural Plan with the Regional Solutions Programme: a more flexible funding programme that provided precisely the type of support for local industry development that the farmer required (Herbert-Cheshire 2003: 468). While the relationship between the activities of the East Warmington farmer and the change in government policy was not a matter of cause and effect, the establishment of various chains of association between these events certainly contributed towards the re-formulation of state objectives.

Conclusion

The inclusion of new, non-state, actors into networks of governing has altered the institutional landscape of agricultural regulation and, with it, the on-farm practices of agricultural producers. Frequently subjected to the demands of global transnational corporations, and of a state that has removed most of the protection once offered, primary producers are forced to compete for market share under terms that are set by, and favour, global capital. In this sense, power may be viewed as having been shifted 'upwards' to transnational corporations which today are, ostensibly, so powerful that nation states have considerable difficulty regulating their activities. At the same time, the emergence of discourses and practices of self-help and state–community partnerships in government policy has prompted some authors to see power as having been devolved 'downwards' to the local level. Together, these shifts have led to a broader debate about the vertical rescaling of state power as the state becomes both too small to manage capitalism in its global form and too large to respond effectively to various changes taking place at the local level (Webb and Collis 2000). Rather than contribute to these somewhat limited debates, this chapter has focused its attention upon the horizontal reconfiguration of state power and the emergence of new governmentalities of rule that are exercised 'beyond the state'. From this perspective, it is argued that the nation state has not so much 'lost' power to local or global actors as has found new, advanced liberal ways of governing through a functioning network of state and non-state actors.

Where this chapter departs from existing accounts of the new governmentalities of rule is by considering not only how state agencies govern through networks of association, but also how 'local' non-state actors are equally able to enrol others into an actor network in an attempt to advance their own goals and objectives, and/or to displace those of the state. The brief examples demonstrate the possibility for contestation and change that arise as a result of collective action. New forms of governing, therefore, cannot be understood simply as a new, advanced liberal mode of rule exercise by political authorities in an era of a 'hollowed-out' nation state. Instead, they create new actor spaces among global, national and local actors in which challenges to the state may be posed, and alternatives to the current neoliberal-inspired globalization path can be articulated. The extent to which these often isolated, and sometimes one-off, instances of local power can be consolidated into a larger, more stable network of resistance similar to the rural protest movements witnessed in Europe in recent years (Woods 2003) is not clear at this stage. Nevertheless, for so long as such challenges continue, the likelihood remains that political authorities will be forced to consider the impacts of their actions upon agricultural producers and their associated rural communities.

Notes

1 One notable exception is the work of O'Malley (1996).
2 See the 'actor-oriented' work of Norman Long and colleagues for an exception to this trend. For example: Long and Long (1992) and Long (2001).

References

Almås, R. and Lawrence, G. (eds) (2003) *Globalization, Localization and Sustainable Livelihoods*, Aldershot: Ashgate.

Barry, A., Osborne, T. and Rose, N. (1996) 'Introduction', in A. Barry, T. Osborne and N. Rose (eds) *Foucault and Political Reason: Liberalism, Neoliberalism and Rationalities of Government*, Chicago, IL: University of Chicago Press.

Brenner, N. (1998) 'Global cities, glocal states: global city formation and state territorial restructuring in contemporary Europe', *Review of International Political Economy*, 5, 1: 1–37.

Burch, D., Rickson, R. and Lawrence, G. (eds) (1996) *Globalization and Agri-food Restructuring: Perspectives from the Australasia Region*, Aldershot: Ashgate.

Callon, M. (1986) 'Some elements of a sociology of translation: domestication of the scallops and the fishermen of St. Brieue Bay', in J. Law (ed.) *Power, Action and Belief: A New Sociology of Knowledge?*, London: Routledge & Kegan Paul.

Callon, M. and Latour, B. (1981) 'Unscrewing the big Leviathan: how actors macrostructure reality and how sociologists help them to do so', in K. Knorr-Cetina and A. Cicourel (eds) *Advances in Social Theory and Methodology: Towards an Integration of Micro- and Macro-sociologies*, Boston, MA: Routledge & Kegan Paul.

Camilleri, J. and Falk, J. (1992) *The End of Sovereignty? The Politics of a Shrinking and Fragmented World*, Cheltenham: Edward Elgar.

Clark, J. and Murdoch, J. (1997) 'Local knowledge and the precarious extension of scientific networks: a reflection on three case studies', *Sociologia Ruralis*, 37, 1: 38–60.

Cooper, D. (1994) 'Productive, relational and everywhere? Conceptualising power and resistance within Foucauldian feminism', *Sociology*, 28, 2: 435–454.

Counsell, D. and Haughton, G. (2003) 'Regional planning tensions: planning for economic growth and sustainable development in two contrasting English regions', *Environment and Planning C: Government and Policy*, 21, 2: 225–239.

Cox, K. (1998) 'Spaces of dependence, scales of engagement and the politics of scale, or: looking for local politics', *Political Geography*, 17, 1: 1–23.

Dean, M. (1994) *Critical and Effective Histories: Foucault's Methods and Historical Sociology*, London: Routledge.

Donaldson, A., Lowe, P. and Ward, N. (2002) 'Virus–crisis–institutional change: The foot and mouth actor network and the governance of rural affairs in the UK', *Sociologia Ruralis*, 42, 3: 201–214.

Edwards, B., Goodwin, M., Pemberton, S. and Woods, M. (2001) 'Partnerships, power and scale in rural governance', *Environment and Planning C: Government and Policy*, 19, 2: 289–310.

Foucault, M. (1978) *The History of Sexuality Volume 1: An Introduction*, London: Penguin.

Gray, I. and Lawrence, G. (2001) *A Future for Regional Australia: Escaping Global Misfortune*, Cambridge: Cambridge University Press.

Green, R. and Wilson, A. (2001) 'Global facts and fictions: industry', in C. Sheil (ed.) *Globalization: Australian Impacts*, Sydney: UNSW Press.

Halpin, D. (2003) 'The collective political actions of the Australian farming and rural communities: putting farm interest groups in context', *Rural Society*, 13, 2: 138–156.

Heffernan, W. (1999) *Consolidation in the Food and Agriculture System*, Denver, CO: Report to the National Farmers' Union.

Held, D. (1991) 'Democracy, the nation-state and the global system', in D. Held (ed.) *Political Theory Today*, Cambridge: Polity Press.

Herbert-Cheshire, L. (2000) 'Contemporary strategies for rural community development in Australia: a governmentality perspective', *Journal of Rural Studies*, 16, 2: 203–215.

Herbert-Cheshire, L. (2003) 'Translating policy: power and action in Australia's country towns', *Sociologia Ruralis*, 43, 4: 54–73.

Higgins, V. (2002) 'Self-reliant citizens and targeted populations: the case of Australian agriculture in the 1990s', *ARENA Journal*, 19: 161–177.

Jones, M. and MacLeod, G. (1999) 'Towards a regional renaissance? Reconfiguring and rescaling England's economic governance', *Transactions of the Institute of British Geographers*, 24, 3: 295–313.

Kelsey, J. (1995) *Economic Fundamentalism: The New Zealand Experiment – A World Model for Structural Adjustment?*, London: Pluto Press.

Latour, B. (1986) 'The powers of association', in J. Law (ed.) *Power, Action and Belief: A New Sociology of Knowledge?*, London: Routledge.

Latour, B. (1987) *Science in Action: How to Follow Scientists and Engineers through Society*, Cambridge, MA: Harvard University Press.

Law, J. (1992) 'Notes on the theory of the actor-network: ordering, strategy, and heterogeneity', *Systems Practice*, 5, 4: 379–393.

Lawrence, G. (1987) *Capitalism and the Countryside: The Rural Crisis in Australia*, Sydney: Pluto Press.

Lockie, S. (1999) 'The state, rural environments and globalization: "Action at a distance" via the Australian Landcare program', *Environment and Planning A*, 31, 4: 597–611.

Lockie, S. and Kitto, S. (2000) 'Beyond the farm gate: production–consumption networks and agri-food research', *Sociologia Ruralis*, 40, 1: 3–19.

Long, N. (ed.) (2001) *Development Sociology: Actor Perspectives*, London: Routledge.

Long, N. and Long, A. (eds) (1992) *Battlefields of Knowledge: The Interlocking of Theory and Practice in Social Research and Development*, London: Routledge.

McKenna, M. (2000) 'Can rural voices effect rural choices? Contesting deregulation in New Zealand's Apple Industry', *Sociologia Ruralis*, 40, 3: 366–383.

McKenna, M., Roche, M. and Le Heron, R. (1999) 'HJ Heinz and global gardens: creating quality, leveraging localities', *International Journal of the Sociology of Agriculture and Food*, 8: 35–51.

McMichael, P. (2003) 'The power of food', in R. Almås and G. Lawrence (eds) *Globalization, Localization and Sustainable Livelihoods*, Aldershot: Ashgate.

McMichael, P. and Lawrence, G. (2001) 'Globalising agriculture: structures of constraint for Australian farming', in S. Lockie and L. Bourke (eds) *Rurality Bites: the Social and Environmental Transformation of Rural Australia*, Sydney: Pluto Press.

Mabbett, J. and Carter, I. (1999) 'Contract farming in the New Zealand wine industry: an example of real subsumption', in D. Burch, J. Goss and G. Lawrence (eds) *Restructuring Global and Regional Agricultures: Transformations in Australasian Agrifood Economies and Spaces*, Aldershot: Ashgate.

Martin, P. (1997) 'The constitution of power in Landcare: a post-structuralist perspective with modernist undertones', in S. Lockie and F. Vanclay (eds) *Critical Landcare*, Wagga Wagga, NSW: Centre for Rural Social Research Key Papers No. 5, Charles Sturt University.

Miller, P. and Rose, N. (1990) 'Governing economic life', *Economy and Society*, 19, 1: 1–31.

Murdoch, J. (2000) 'Networks: a new paradigm of rural development?', *Journal of Rural Studies*, 16, 4: 407–419.

Murdoch, J. and Marsden, T. (1995) 'The spatialization of politics: local and national actor-spaces in environmental conflict', *Transactions of the Institute of British Geographers*, 20, 3: 368–380.

O'Malley, P. (1996) 'Indigenous governance', *Economy and Society*, 25, 3: 310–326.

Robertson, G. (1997) 'Managing the environment for profit', in Australian Bureau of Agriculture and Resource Economics, Outlook 97, Proceedings of the National Agricultural and Resources Outlook Conference, Canberra: ABARE.

Robertson, R. (1992) *Globalization: Social Theory and Global Culture*, London: Sage.

Rose, N. (1993) 'Government, authority and expertise in advanced liberalism', *Economy and Society*, 22, 3: 283–299.

Rose, N. (1996) 'Governing "advanced" liberal democracies', in A. Barry, T. Osborne and N. Rose (eds) *Foucault and Political Reason: Liberalism, Neoliberalism and Rationalities of Government*, Chicago, IL: University of Chicago Press.

Rose, N. and Miller, P. (1992) 'Political power beyond the state: problematics of government', *British Journal of Sociology*, 42, 2: 173–205.

Stoker, G. (1998) 'Governance as theory: five propositions', *International Social Science Journal*, 155: 17–28.

Teubal, M. and Rodriquez, J. (2003) 'Globalization and agro-food systems in Argentina', in R. Almås and G. Lawrence (eds) *Globalization, Localization and Sustainable Livelihoods*, Aldershot: Ashgate.

Webb, D. and Collis, C. (2000) 'Regional Development Agencies and the "new regionalism" in England', *Regional Studies*, 34, 9: 857–873.

Woods, M. (1997) 'Researching rural conflicts: hunting, local politics and actor-networks', *Journal of Rural Studies*, 14, 3: 321–340.

Woods, M. (2003) 'Deconstructing rural protest: the emergence of a new social movement', *Journal of Rural Studies*, 19, 3: 309–325.

Woods, M. (2004) 'Political articulation: the modalities of new critical politics of rural citizenship', in P. Cloke, T. Marsden and P. Mooney (eds) *Sage Handbook of Rural Studies*, London: Sage.

4 Rural development and agri-food governance in Europe
Tracing the development of alternatives

Terry Marsden and Roberta Sonnino

Introduction: tracing the battleground of agri-food in Europe

In some ways, rural space within Europe has become a 'battlefield' of knowledge, authority and regulation fought around different definitions of agri-food. The outcome of this battle will shape not only the 'quality' of food, but also the rural space itself – its resource potentialities, its governing and its sustainability. Involved in this struggle are three different social, political, scientific and economic paradigms which combine and, at the same time, compete for primacy in the policy development process (Marsden 2003: 4). First, the *agro-industrial paradigm*, associated with the globalized production of standardized food commodities and with recent political attempts to 'deregulate' international markets. Informed by a neo-liberal 'virtual' logic of scale and specialization which ties agri-food into an industrial dynamic, and privileges national and international perspectives, this continues to be by far the most powerful development paradigm governing the production of agri-food in Europe. Second, the *post-productivist paradigm*, based on a perception of rural areas as consumption spaces to be exploited not by industrial capital but by the urban and ex-urban populations. Emerging in the past decade or so, the post-productivist model challenges the agro-industrial paradigm through an alternative emphasis on the local environment and environmental protection for its own sake. Yet it shares with it the marginalization of nature – which in the agro-industrial model takes place through the production process, but in the post-productivist model it occurs through urban consumption of the countryside and the marginalization of the agricultural economy. Third, the newly emerging (and more contentious) *sustainable rural development paradigm*, which redefines nature by emphasizing food production and agro-ecology in the context of a more multi-functional rural context. In contrast with the other two dynamics, this model emerges at the regional rather than national or local level, and it stresses the 'embeddedness' of food chains – or, more specifically, their association with highly contested notions of 'place', 'nature' and 'quality'.

Associated with this sustainable rural development model, new grassroots and alternative forms for the governing of food are emerging (see Figure 4.1). On the one hand, these are establishing themselves in commercial and ideological opposition to the dominant modes of governing and regulatory systems and, more generally, to the backdrop of globalization. Through rationalization of production sites, techniques and market operations, global competition offers some competitive advantages. However, the process itself tends to distribute costs and benefits unevenly across different spatial, temporal and social domains. In this context, local forms of development, including local food systems, potentially provide an effective counterforce and spatial platform against the economic, political and social vulnerability of communities that are not located on the benefit side of the global logistics scale (Marsden and Smith in press).

On the other hand, the emerging alternative forms for the governing of food are receiving some support from current European policies, particularly the 'Second-Pillar' Rural Development Regulation (RDR), recently introduced in the context of Common Agricultural Policy (CAP) reform, and the opportunities to gain the status of derived geographical origin (PDO/PGI) for specific local products (Parrott *et al.* 2002). By creating hitherto unavailable opportunities for regions to design sustainable rural development strategies attuned to their own needs, the RDR could potentially enhance the role of sub-national actors in influencing the implementation of policy.

In this chapter, we analyse the implications of these increasingly competing domains of space and governing to assess how a more sustainable rural development model might take hold. Specifically, we ask: *To what extent are the new and alternative frameworks of governing stimulating a change of emphasis from the established 'agro-industrial' and 'post-productivist' models towards a more specifically (and sustainable) agro-food-based rural development model? That is, how is the battlefield playing itself out over time and space in Europe?*

More generally, and in relation to progressing theorizations of agricultural governance, we argue that the *relocalization of agro-food* is far more than just a reaction to globalization. In fact, it can reactivate rural space as a live agent in the shaping of the 'competitive spaces' between 'conventional' and 'alternative' food sectors (see Figure 4.1). Methodologically, this calls for scholars to analyse comparatively and ground empirically the 'battlefield' model we identify – i.e. to compare and contrast (both nationally and internationally) different regions and localities affected by these new dynamics. To progress this approach, we will first explore the nature and theoretical implications of the competing spatial and regulatory domains that have developed in Europe, and then analyse empirical evidence on the recent governing and agro-food developments in the UK. For our analysis, we have selected as case studies Wales and South West England. While both regions have been among the most proactive in the UK in the development of relocalized food systems, they differ substantially in terms of governing. This creates an opportunity to analyse comparatively the general relationships between emerging frameworks of governing and regional innovation, and

Type of spatial relationships	Delocalization Conventional agri-food	Relocalization Alternative agri-food
Producer relations	Intensive production 'lock-in'; declining farm prices and bulk input suppliers to corporate processors/retailers.	Emphasis on 'quality'; producers finding strategies to capture value-added; new producer associations; new socio-technical spatial niches developing.
Consumer relations	Absence of spatial reference of product; no encouragement to understand food origin; spaceless products.	Variable consumer knowledge of place, production, product and the spatial conditions of production; from face-to-face to at-a-distance purchasing.
Processing and retailing	Traceable but privately regulated systems of processing and retailing; not transparent; standardized vs. other than spatialized products.	Local/regional processing and retailing outlets; highly variable, traceable and transparent; spatially referenced and designed qualities.
Institutional frameworks	Highly bureaucratized public and private regulation; hygienic model reinforcing standardization; national CAP support (Pillar I).	Regional development and local authority facilitation in new network and infrastructure building; local and regional CAP support (Pillar II).
Associational frameworks	Highly technocratic – at a distance – relationships; commercial/aspatial relationships; lack of trust or local knowledge.	Relational, trust-based, local and regionally grounded; network- rather than linear-based; competitive but sometimes collaborative.

← CHANGING COMPETITIVE SPATIAL BOUNDARIES →

Figure 4.1 Rural space as competitive space and the 'battleground' between the conventional and alternative agri-food sectors

the multi-level process of regional and institutional building of local food networks.

The regionalization of rurality in Europe and the UK: an overview

The past fifty years have witnessed a radical shift in the political perception of the UK countryside and in the forms of governing associated with it. As Murdoch and co-workers (2003) explain, post-war economic development demanded an agriculture that played a central role in national reconstruction. This requirement shaped a political discourse centred upon the 'national farm', or, in simple terms, the idea that all agricultural areas must contribute to the greater national 'good'. For decades, this doctrine promoted the development of a strong and unified national policy framework, guarded by an extensive and centralized administrative apparatus revolving around the Ministry of Agriculture and the National Farmers' Union. This discouraged local and regional variations, and tied all regions of the UK to the goal of contributing to the national productivist regime. Entry to the EU, initially at least, did not change this regime significantly.

However, by the early 1980s the dominant political discourse on agriculture began to be profoundly questioned. In fact, on the one hand, it was clear that agriculture had ceased to be the main economic activity in rural areas. On the other hand, it became evident that farming (and the operation of the agro-industrial model) had caused major environmental damage to the countryside. These two factors, coupled with the ever-increasing costs of agricultural support and with a series of food and farming crises, provoked a significant rethinking of agricultural policy both at the national and the European level.

In the UK, one significant outcome of this process was the publication of the Rural White Papers in the mid-1990s, conceived to spell out the broad range of state initiatives in operation in rural areas. With the White Papers, the concepts of 'regional' and 'local' began to emerge in the UK's political discourse. In fact, the Papers were produced on a devolved (England, Scotland and Wales), rather than UK-wide, basis – a strategy which tended to highlight more regional differences in rural conditions. In contrast, underlying the White Papers was the idea that rural policy should harness local variety in line with broader policy goals (Murdoch *et al.* 2003: 49).

After the election of a Labour government in 1997, the political attention in the UK switched further to the regional level. One of the new government's early acts was to allow devolution campaigns in Scotland and Wales which led to the formation in 1999 of a Scottish Parliament and of the Welsh Assembly, both carrying an enhanced responsibility for rural policy. In England, the most significant initiative in this respect was the establishment of Regional Development Agencies (RDAs), which assumed the responsibility of providing coordinated regional economic development and regeneration (Winter 2003: 11). In terms of rural policy, the RDAs are now in charge of promoting integrated development, coordinating the work of public agencies and delivering a range of services that respond to the needs of rural residents (Murdoch *et al.* 2003: 50).

The Foot and Mouth crisis in 2001 made an important contribution to the consolidation of the UK's regional approach to rural policy in two fundamental ways. First, as Winter (2003: 8) explains, during the crisis, the Ministry of Agriculture was so absorbed in dealing with the disease itself that it was left with little spare capacity to address the devastating socio-economic effects of the crisis (most significantly on rural tourist businesses). The RDAs and local authorities stepped into this policy vacuum, and, while dealing with issues of economic recovery, they also strengthened their role in the emerging regional system of rural governing. Second, the crisis brought local and regional governing together around a new agenda for sustainability (Winter 2003: 11). This was presented for the first time in the *Report on the Future of Farming and Food*, known as the 'Curry Report' (2002), which attempted to chart a course out of the crisis.

In advocating local food as 'one of the greatest opportunities for farmers to add value and retain a bigger slice of retail value', the Report emphasizes how:

> local food markets could deliver on all aspects of sustainable development – economic (by providing producers with a profitable route to market), environmental (by cutting down on the pollution associated with food transportation, and by interesting consumers in how the land around them is farmed) and social (by encouraging a sense of community between buyer and seller, town and country).
>
> (Policy Commission 2002: 119)

To overcome the existing barriers to the development of local food markets, the Curry Report encourages the RDAs 'to devise a regional food component to their regional economic strategies' (Policy Commission 2002: 46) and, more specifically, to consider how to overcome problems of distribution and availability, and to promote the networking and planning that are necessary for the development of local food initiatives (Policy Commission 2002: 45).

With specific regard to England, the process of rural regionalization has recently been boosted by Lord Haskins' review of the delivery of rural policies. Appointed in November 2002 by the Minister of State for Rural Affairs to improve rural delivery (particularly in DEFRA), Haskins has recommended devolution. As stated in his Report:

> I would like to see rural delivery in England becoming much more decentralised than it is, with key decisions being taken at regional and local levels. This is where services can most effectively address public need and where deliverers can be held more clearly to account.
>
> (Haskins 2003: 8)

More specifically, the Report claims that 'behind the picture of bureaucratic complexity and customer confusion in rural delivery arrangements' lies a 'long history of . . . a national approach to solving problems that are largely local in nature' (Haskins 2003: 12). Such an approach, according to Haskins, has become 'over-centralized and unmanageable', and cannot ensure 'clear accountability' and 'responsiveness to need'. To achieve these goals, there is a need for an 'extensive devolution of policy delivery to regional and local networks' (Haskins 2003: 13) such as the RDAs, local authority and the voluntary and community sector (Haskins 2003: 43).

The emergence of 'regionalized ruralities' (Murdoch *et al.* 2003: 10) in the UK reflects (and it is promoted by) the occurrence of a similar process at the EU level. When, in the early 1980s, the problems of overproduction and environmental degradation associated with the operation of the CAP became evident throughout Europe, the Commission began to rethink its agricultural policy. This started a series of political reforms that culminated with the introduction of the RDR, or the Second Pillar of the CAP (the first being commodity support).

In general terms, the RDR has two major impacts on European rural policy and governing. First, it creates an institutional framework for reorienting the direction of European policy from agriculture to rural development. Second, it is promoting a regional approach to programming the CAP. In the UK, Scotland and Wales are administering their Rural Development Plans (RDPs) through their devolved governments, whereas the English Plan has nine regional chapters. As Murdoch and co-workers write (2003: 38), 'it would seem that as rural development becomes a more important aspect of agricultural policy . . . that policy is likely to become more regional in nature'.

The recent reform of the CAP, agreed by the EU Agriculture Ministers in June 2003, allows scope for devolution also within Pillar 1. In this respect, the cornerstone of the reform is the introduction of the Single Farm Payment (SFP), which is independent from production and linked to complying with environmental, food safety and animal welfare standards ('cross-compliance'). Under the reformed CAP, the SFP was implemented at the regional level, starting from January 2005, following two possible modalities. Regions have the option to adopt an 'area-based' approach, whereby the SFP will include all farmers, regardless of whether or not they are currently receiving CAP support, or to use an 'historical' approach that keeps the SFP linked to historical CAP entitlement. While Wales has opted for an historical approach, which better responds to the needs of a narrow-based farming sector dominated by livestock, England has chosen, after endless discussions and assessments, to implement an area-based SFP. The English approach is therefore seen as a more radical and commercially based approach to CAP reform which reflects the larger scale of farms in lowland England and the need to encompass the intensive pig, poultry and horticultural sectors, which were previously outside the CAP subsidy system. In Wales, by contrast, the 'historic' method for Pillar 1 has been adopted in order to give more support for the smaller family farm.

In short, the system of rural governing in both England and Wales has been subject to at least four different, if related, processes: political devolution, a steady process of social and economic regionalism, a significant national rural policy review, and a decade of wider EU CAP reform. Ostensibly and theoretically, the convergence of these trends opens up new contingencies for the relocalization and regionalization of agro-food, and it presents a significant challenge to the agro-industrial conventional system. In the following two sections, we will explore how these processes have been playing out in South West England and in the now devolved case of Wales.

Regional contingency 1: The relocalization of governing and food in South West England

When asked to comment on the new regional approach to the CAP reform, a senior development adviser from the Food, Farming and Rural Development Sector of the Government Office for the South West said:

> I remember one meeting where people wanted to go for the historic-based scheme because they felt funding would be inappropriately diverted away from the areas of greatest need, particularly the livestock sector. We got three main moors in this region, where you got a lot of extensive hill livestock, which has a very limited market, the income is very low without support. . . . One of the arguments for supporting the livestock sector is that the sector is important in managing the environment.

Besides illustrating the typical tension between conventional and alternative farming enhanced by the recent CAP reform, these words point to two fundamental characteristics of the rural South West. First, its propensity to actively re-elaborate rural policies superimposed upon the region; second, a convergence of interests between environmentalism and rural development that lies at the foundation of its local identity.

In contrast with Wales, this region witnessed a widespread increase in both population and employment during the 1990s. The scale of the development pressures resulting from these trends stimulated the growth of environmentalist groups, especially among newcomers. In some ways, these groups have forged an alliance with the rural development network. In fact, while environmentalists recognize that the conservation of the countryside relies on a viable and sympathetic agricultural industry, the rural development network accepts that farmers have a social responsibility as guardians of the countryside.

In other ways, however, there is a tension in the South West between the environmentalist and the rural development agendas. Environmentalists attempt to bring rural areas within discourses of environmental valuation through the assertion of ecological conventions. Farming and rural development interests, in turn, assert conventions associated with growth and market demand in their concerns about options and choices available to the numerous rural families who seek to sustain their livelihoods (Murdoch *et al.* 2003: 101). We have typified these socio-political relations as an example of a *contested countryside* characterized by mutual suspicion, with the environmentalists fearing that intensive farming would damage the environment, and rural development and farming interests suspecting that environmentalism may constrain the opportunities of farmers and developers to pursue new commercial possibilities.

The official political discourse in the South West purports to reconcile the regional rural development and environmental interests – or, as the rural development adviser put it during the interview, 'to encourage integration of agri-environmental schemes with rural development schemes'. The regional RDP is a product of this political effort. In fact, its aspiration is 'to enable the South West rural communities to retain and strengthen their cultural distinctiveness, economic viability and quality of life through integrated rural development which conserves the special character and diversity of the region's environmental assets' (DEFRA 2000: 139). Under the assumption that, by promoting environmental diversity, sustainable or integrated rural development also contributes to economic development and to the social well-being of the region in the allocation of funding, the plan strongly emphasizes agri-environmental measures and afforestation of farmland.

From a practical perspective, an analysis of how the discourse attempting to integrate environmentalism and rural development is implemented shows that in the South West there is very little responsiveness on the ground to political plans and strategies that are perceived to be coming from above. Our empirical work in the region revealed a generalized lack of interest (and even knowledge) concerning the RDP, which the rural development adviser whom we interviewed explained in the following terms:

> Basically the RDP came from above, you got this collection of ten schemes, it was European, and then the British government decided on what the priorities were, how was it going to implement that within England, then there were the regional chapters, but basically we had to work within the structure of those schemes.

A representative from DEFRA confirmed that the RDP in England has not so far promoted regional autonomy in that 'the bulk of the schemes tend to be directed from the centre'. However, he foresees a much greater degree of 'regional discretion' in the future, for two reasons. In his words:

> One, there's going to be a review of the Second Pillar Regulation itself in Brussels and the pressures from everywhere are for a CAP free of

bureaucracy. . . . But secondly, when we'll come to decide how to implement it here, the regional–centre debate has moved on quite a long way from 2000.

Parallel to the official political discourse coming from London, there is another discourse in the South West that is used to establish (and market) a regional identity. Inspired by the aftermath of the Foot and Mouth crisis and by the recommendations provided by the Curry Report, such discourse relocalizes food as a potential means to reconcile the environmentalist tradition of the region with its rural development interests.

The Foot and Mouth crisis and the Curry Report had two major impacts on the food and rural agenda of the South West. First, they promoted a process of regionalization with regard to food and farming. This process had at the forefront an especially active RDA. As Winter (2003: 8) points out, in the South West the RDA promptly occupied the policy vacuum created by the new necessity of addressing issues of economic recovery after the Foot and Mouth crisis. In fact, even before the end of the investigation that led to the Curry Report, the RDA commissioned research on the economic impacts of the disease, organized an economic and social summit on it, supported the establishment of a new Chamber of Rural Enterprise, funded some early recovery projects, and extended the Market and Coastal Town initiative to cover some of the hardest hit market towns in the region.

Second, the aftermath of the Foot and Mouth crisis has also contributed to shape a vision of local food systems that fits well with the South West's attempt to integrate rural development and environmentalist interests. In discussing why the RDP has not been effective in guiding the regional rural policy, the adviser said:

> The RDP came before Foot and Mouth; Foot and Mouth really focused everybody's mind on the importance of the food and farming industry and the way it interacts with so many other industries, with landscape management and so on.

Indeed, the plan is the product of a cooperative effort that brought together a number of different stakeholders both from the rural development network and the environmentalist network. Specifically, the first interest group includes representatives from the South West Rural Affairs Forum and DEFRA rural development service, whereas the environmental network is represented by members of the Countryside Agency and the South West Regional Environmental Network.

The bottom-up and regionalized approach to the development of the agri-food sector that has characterized the policy of the South West in recent years has been consolidated through the establishment of an umbrella trade organization for the food and drink industry called South West Food and Drink (SWFD), which receives new funding from the RDA. One of the project managers of the organization explained during an interview that there were two main reasons behind the decision to form SWFD: the need to avoid 'inefficient devolution of publicly funded money' and, again, the desire to promote and coordinate cooperation among different agencies that support the industry.

Informed by the vision that 'the South West of England is recognised as the region of excellence for food and drink', SWFD has a fundamental mission: 'to create a sustainable regional food supply industry that delivers long-term profitability and prosperity for the South West.' Here, food serves again as a link between the environmental assets of the region and its opportunities for economic development. Such initiatives represent new rural development processes based upon an integration of rural development and environmentalist concerns. In this sense, rural development – based here upon a range of coordinated agro-food initiatives at the regional and local levels – is beginning to challenge the more established models of agro-industrial convention and post-productivism. Significantly, it is encouraged to do so through the operation of regional bodies and organizations that weld together the formerly contested politics of agricultural productivism and ex-urban post-productivism.

To summarize, the South West has successfully built on the opportunities for regionalization opened up by the devolutionary tendencies that have developed in the UK over the past few years. Through the adoption of a bottom-up approach to rural and agri-food development, the region has created a regionally based cooperative network of local stakeholders who are promoting a new vision of 'distinctive, regionally associated and environmentally friendly' food products. At one level, this vision reconciles the conflicting environmental and rural development interests that have developed in the South West over the past decade, thereby strengthening the identity of the region. At another level, the new vision of food promoted by organizations such as SWFD is also marketing the regional identity, turning the South West into the 'region of excellence' with regard to food and drink.

Regional contingency 2: Reshaping rural and regional governing in Wales

The agrarian political history of rural Wales was shaped by a complex interrelationship of regional 'marginalization' and 'clientelism'. Welsh farmers were traditionally integrated into a productivist 'economy of scale' model that was part of the national agricultural strategy. Given the upland nature of much of its agriculture, Wales came to be considered almost immediately as a 'problem region', one in need of 'agricultural adjustment'. The small size of its holdings made it extremely difficult for farmers to mechanize and modernize. Socially, these difficulties were compounded by rural depopulation. The most evident result of these pre-conditions was the emergence of a form of 'arrested marginalization' that, by requiring strong state intervention and support, facilitated throughout the post-war period the development of a 'clientelistic countryside' – one which is 'beholden to the State for its definition, its dynamic and its distributive powers' (Marsden 2003: 125).

After the mid-1950s, when the need for a distinctive Welsh rural politics began to take hold (see Murdoch 1992), such development took the shape of a steady process of 'clientelistic devolution', with the marginal regional rurality of Wales guarded by an increasingly devolved state apparatus. For decades, this complex form of rural clientelism remained dominated by the agricultural interests of a fairly stable policy community involving the Welsh Office Agricultural Department (WOAD) and the two farmer representative organizations: the National Farmer Union and the Farmer Union of Wales (Murdoch 1992).

The year 1996 was a turning point for the clientelistic countryside of Wales. Early that year, the evidence of the possibility for transference of the BSE to humans, combined with the centralization of the dairy processing industry and the rising costs of transport, seriously affected the Welsh livestock market, causing a dramatic decline in farm gate prices. This crisis led to two significant developments: it promoted a 'supply-chain approach' to the farm crisis among both farmers and government authorities, and it significantly broadened the nature of rural clientelism. With regard to the emergence of a 'supply-chain approach', as the farmer unions struggled to represent their members with regard to the powerful supermarkets, Welsh farmers took action in their own hands. By blockading imports of Tesco hamburgers from Ireland, for the first time they intervened at a different level of the food supply chain. At the same time, newly established structures of governing began to replace the old productivist and compensatory system for alleviating problems in the farm sector with a supply-chain remedy that brought food into the political arena.

A Food Strategy for Wales, produced in 1996 by an advisory group that included politicians as well as representatives from producers, consumers and farmer organizations, is one of the earliest official documents to be based on the newly developed 'supply-chain approach'. In fact, the strategy rests on three pillars that embrace all actors in the food chain. These include: (1) supporting Welsh *products* in terms of 'quality, range and availability'; (2) developing *markets* for Welsh products; (3) creating technical and business support *services* for the food industry (WOAD 1996: II–III). Three years later, the establishment of the Agri-food Partnership also brought together a variety of private and public sector actors operating across the whole supply chain 'to provide focus and a "joined-up" approach for the development of the agrifood industry in Wales' (WDA 2004: 1). This reflected a new alliance between the WOAD and the Welsh Development Agency (WDA) which, for the first time, became involved in developing agri-food policy. In this new system, three sector groups were established to identify targets and develop action plans for the red meat, dairy and organic sectors. Again, these plans promote the adoption of a 'supply-chain approach' to the farm crisis.[1]

In the two years which followed the publication of the three action plans, the Partnership took a series of initiatives that continued to target different aspects of the food chain. Ranging from the development of school programmes to encourage milk consumption among children to the establishment of Farming Connect (a coordinated support programme for producers), from marketing initiatives to secure contracts for Welsh lamb and beef in Safeway stores to financial investment in the organic food processing sector (Agri-food Partnership 2001: 10–11), such initiatives have formed 'an integrated approach towards the development of an industry that incorporates primary farming production, processing, wholesaling, retailing and the marketing and distribution functions' (Agri-food Partnership 2001: 8).

During an interview, a top representative of the Agri-food Partnership defined 'supply-chain linkages' as their 'absolutely paramount priority'. At a broader level, the Director for Agriculture and Rural Affairs said that while there has been significant progress in the work on marketing and branding for the red meat sector in Wales, more needs to be done 'in the middle', or on the processing side. To him, this requires action in the context of the CAP reform:

> What I'm struggling with a little bit is . . . we got Farming Connect, and that's working okay, we got the Agri-food Partnership, we got the various strategies for lamb and beef, dairy, etc. . . . but how can we use both Pillar 1 and Pillar 2 money differently . . . to better encourage local cooperation and local procurement?

As a result of the development of the 'supply-chain approach', the nature of clientelism in rural Wales has changed significantly in recent years. State dependency is now no longer exclusively associated with agricultural productivism or the narrow corporatist relationships allied with the WOAD and the two farmers' unions. Rather, there is now a much broader 'rural' clientelism related to the need for state support and initiatives across a wide range of socio-economic activities concerning rural areas.

Partly associated with the attempt to use government agencies to unlock more funding for rural development coming from Brussels, this new type of clientelism encourages a system of multi-level governing whereby the articulation of networks among the local, national and European policy becomes the norm. Such a system has both vertical and horizontal dimensions. In fact, it responds to the need of obtaining blocks of funding from Cardiff or Brussels or both, and it addresses the necessity of building the types of effective partnerships and cooperative arrangements that are necessary to enter the competitions for funding (Marsden 2003: 132).[2] In other words, the multi-level structures of governing that have appeared recently 'are vertically interlinked through operating at different spatial scales (European, UK, Wales, sub-regional, local authority), but they also have to engage with a wide variety of external actors and networks (both state and non-state) at each of these levels' (Marsden 2003: 134).

In this context, the State is no longer just a formulator and deliverer of policy. It is an orchestrator of agricultural, environmental and rural development networks formed by state agencies that can only act in relation to others. As a spatially-based, rather than sectoral agency, the WDA acts as a major coordinator of rural activities, and it does so with a clear awareness of the need for a networked approach.

In the context of a continuing and deepening agricultural crisis that has fuelled the political need to develop separate strategies for different UK regions, Wales has developed its own specific model of 'facilitative clientelistic rural governance' (Marsden 2003: 136). Ideally, this is promoting the distinctiveness of Welsh rural life and small-scale farming (i.e. the family farm) through the need to compete on the basis of social, economic and environmental 'quality' food criteria, rather than on quantity and price. As stated in *Farming for the Future* (Government of the National Assembly for Wales 2001: 9), 'if Welsh farming and food processing continue to try to compete on price alone in basic food commodity markets, the result will be a strong continuation of the long-term trends which have been eroding the pattern of family farm' – a pattern that, as stated elsewhere in this document (2001: 7), defines the character and sense of identity of Welsh rural society. The alternative envisioned in *Farming for the Future* is 'to move as far as possible along the spectrum towards competing less on price and more on quality' (Government of the National Assembly for Wales 2001: 13). In practice, this means 'developing high quality, value-added, branded products which are aimed, where possible, at more special markets and niche markets' (Government of the National Assembly for Wales 2001: 13).

In the context of this new type of rural clientelism, Welsh rural policies are informed increasingly by the ideals of sustainable development and integrated rural development. These derive far more from a mainland European context than from a specifically British one (see Prodi 2002). The regional RDP also emphasizes economic and environmental sustainability among the strategic priorities for rural Wales. As a result, funding mechanisms (such as EU Structures funding, the rural development regulation and LEADER programmes) are used in different ways and in different parts of rural Wales. The task and continuing challenge of the Welsh Assembly government and the WDA is to match and manage this process in line with the different strategies for agri-food, farming and rural development, so as to obtain as much synergy as possible between the policy frameworks and the different types of rural and agricultural needs.

Agriculture and agri-food are thus re-emphasized and repositioned for their contribution towards achieving environmental and socio-economic sustainability. In this sense, the recent socio-political transition within the rural clientelistic state of Wales is reconstituting agricultural governing, rather than marginalizing it (Marsden 2003: 138). In accordance with the spirit of the CAP reform, this is based on a neo-productivist and multi-functional view of agriculture that has emerged in Wales within the context of a strong agri-environmental policy. Significantly, this approach has been presented explicitly as Welsh both in its essence and its goals by the policy-makers interviewed. In discussing the prospects of CAP reform, a representative from the Director for Agriculture and Rural Affairs referred to the new CAP reforms as 'great opportunities for us not only to make the industry and the countryside more sustainable, but actually to improve our brand, to improve our marketing ability in the UK and the rest of the world'. A member of the National Assembly said:

> We have tried to take a Welsh approach to the Pillar 2 stuff, which is why we have our own Less-Favoured Areas scheme, we got our own agri-environmental scheme . . . and Farming Connect is an example of the way we want to go in terms of helping the industry to adapt. We think this is best suited to Welsh circumstances.

In sum, Wales' strong history of agri-environmental schemes over the past decade has actively supported the transition from productivist to a new type of neo-productivist and multi-functional agriculture that responds to the logic and opportunities presented by the emerging rural development paradigm rather than the conventional agro-industrial model. Indeed, the RDP for Wales identifies sustainable agriculture as a priority for the region. Similarly, *Farming for the Future* promotes a multi-functional type of agriculture that delivers safe and healthy food and non-food products, a visually attractive countryside, and distinctive local food products that support tourism and a positive image of Wales (Government of the National Assembly for Wales 2001: 12).

In short, at least at a theoretical level, Wales has capitalized on the new opportunities for regional distinctiveness offered by the CAP reform and by an increasingly rooted philosophy of devolutionism in the UK. As a leader of the Agri-food Partnership explained during an interview: 'We have been able through devolution to look at what our industry is and where it can go, rather than taking this broad brush approach which fits some.' This process, and the benefits it can have, is acknowledged also by the UK central government. A representative from DEFRA in London commented on the recent changes:

> I think we were very clear, all of us, particularly following the experience of modulation, that the more we could get rid of our mutual interdependence, the better. So we invested quite a lot of effort in the [EU] Council to get wherever we could these revisions that said decisions can be made not just at the member state level but at the regional level as well. . . . What we have achieved is we have made devolution more of a reality.

This analysis has shown that it is indeed through the broader process of devolution that the region has become increasingly more capable of developing an innovative and more endogenous form of rural and agricultural governance. This tends to respond to, and at the same time promote, the new European model of multi-functional agriculture and the emerging sustainable rural development paradigm.

Conclusions: Intensifying competition between the agri-industrial and rural development models

We began this chapter by typifying European rural space as a 'battlefield of knowledge, authority and regulation', fought around different definitions and conventions of agri-food. We have seen in the two regional case studies how the development of an alternative agri-food/rural development paradigm has begun to emerge at the same time that the agro-industrial model has been subject to further economic crisis and public concern. During the recent periods that both case studies document, conventional farm incomes have continued to fall and farmers who are locked into this system have become even more dependent upon the conventional forms of CAP support. In addition, in both regions it was the severe animal health crises (Foot and Mouth in the South West, and the earlier but even more profound BSE problem affecting the beef sectors in Wales), associated with the intensive system, which pressured many producers and food chain actors to 'unlock' themselves from this agri-industrial model. In both cases this also spurred institutional change, since it coincided with the growing devolution and regional agenda by national government and with the more proactive reform of the CAP.

As a result, we may conclude that significant institutional space has opened up, such as to allow and empower the alternative rural development paradigm – at least as it relates to developing more local and regional strategies for endogenous agri-food development, based upon a food chain perspective, rather than the old sectoral and corporatist systems. This does not mean, however, as in many other rural regions of Europe (see Andersson *et al.* 2003; van der Ploeg 2003), that the agro-industrial model has necessarily receded. This model in the UK context has become more retailer-led, coupled with state-based, highly bureaucratic and hygiene-based systems of risk management and assessment. Agencies such as the European Food Safety Agency and the nationally based Food Standards Authority are currently devising systems of traceability and cross-compliance which will further police and intervene in conventional farmers' practices so that risk minimization and standardized systems of control may be attained, supposedly in the consumer's interest. This is likely to increase further the regulatory costs for conventional production, promote economies of regulatory scale, and reinforce *'lock in'* for many producers. For instance, in both regions, but especially in the South West, the size of the conventional dairy herd needed to sustain a farm continues to increase as farm gate prices at best stay static. In the agro-industrial world, the continuity of the traditional farm-based cost–price squeeze is still endemic, and many government officials in DEFRA see the decline in the number of farms as an inevitable feature of 'the market'.

In the South West and Wales, however, this neo-liberal view of agro-food is now significantly challenged, and by various means an alternative and more embedded system of production and supply is being assembled. This is relying upon recasting the web of political and regulatory relations between the EU and the regions, between the UK government and the regions, and between and within the regions themselves. Thus, unlike earlier periods, when farm diversification or rural diversification was considered to be the panacea for the peripheral rural regions, now such bottom-up initiatives not only have their own stronger dynamic; they also have significant institutional support. Neither the South West nor Wales' recent collection of strategy documents suggests that there should be an 'inevitable' reduction in the number of family farms. Quite the contrary, through forging new alliances among rural development, environmental as well as agricultural interests and networks, agri-food is being steadily repositioned as a neo-productivist activity that is again central for broader aspects of sustainable rural development. This cannot be simply left 'to the market' (as some officials in DEFRA suggest).

In part, the new agenda for regional governing has played a crucial role here in fostering these more concerted actions among networks that were hitherto very often in conflict with each other. It has, quite uniquely in the UK sense, forced these networks to reassess, more cooperatively, their economic, social and cultural assets with regard to their territories and it has then stimulated them to rearticulate the quality, value and 'spatial worth' of these alternatives. In this sense, agri-food and broader aspects of rural development start to become significant contributors to the wider social and political process of 'new regionalism'. This is, however, a highly contested and contingent process which is liable to 'backlash' politics, not least on occasions from traditional farmer union interests, but also from central UK government agencies and corporate retailers.

Nevertheless, in South West England and Wales as well as in other European rural regions, we see the beginnings of an 'autonomous rupture' occurring with the central state; and in the Wales case, a promotion of this by the UK state (DEFRA) itself, as it starts to act for English regions as opposed to the UK as a whole. From the point of view of progressing the alternative agri-food and rural development agenda, then, our analysis suggests that for both producers and state agencies, it may be the *degree of detachment*, rather than integration and control, displayed between central and regional actors which becomes a key mechanism for providing new spaces for alternative and more ecologically oriented actions. It would seem that the opportunities for exploiting the regional potentialities of broader European systems of governing can give, at the very least, further impetus for a more viable rural development model based upon re-embedded food chains. But the battle has only just begun.

These unfolding, evolving and highly contingent forms of agri-food governing outlined in this analysis suggest the need for more comparative research particularly within Europe, but also between European and non-European rural regions, as the macro-political and economic battlegrounds – between conventional and alternative agri-food systems – continue. The case analysis here demonstrates how this implicates the central (EU and UK) and regional state; and how innovative forms of regionally based actions and strategies can potentially contribute to a more sustainable form of (regionalized) rural development. Future work on agri-food regulation, therefore, needs to accommodate both a vertical problematization associated with contested and layered multi-level systems of governing on the one hand, and a more fine-grained empirical lateral/embedded analysis of the contested spatial dynamics associated with the agro-industrial, post-productivist and rural development paradigms.

Acknowledgements

This chapter is based upon a research project supported by the Economic and Social Research Council (UK) entitled: 'Going Local? Innovation Strategies and the New Agri-food Paradigm' (ref: Toe50021). The evidence cited in this text derives from in-depth interviews and documentary analysis conducted by the authors and colleagues in the first phase of the research project which sampled key policy officials at the different levels of governing (EU, UK, Wales and South West England). The authors also wish to acknowledge the support of their two partners in the project: Kevin Morgan and Jon Murdoch.

Notes

1 The Strategic Action Plan for the Welsh Lamb and Beef Sector, for example, emphasizes the 'urgent need for all the players in the "traditional" red meat supply sector – producers, auction markets, abattoirs and meat processors – to work much more closely to differentiate their products further and gear up to meet the needs of a rapidly developing consumer marketplace' (Agri-food Partnership 1999a: 3). Similarly, the Strategic Action Plan for the Organic Food Sector encourages action at a number of different levels of the food chain – including providing support for producers, the processing sector, retailers and caterers (Agri-food Partnership 1999b: 18–19).

2 Significantly, *Farming for the Future*, a document produced by a group of experts in 2001 to advise the Minister Carwyn Jones on the direction the industry should take, explicitly identifies a dual political role for the National Assembly: (1) to influence the UK government and the EU so as to secure a trading and subsidy framework for Welsh agriculture; (2) to take action – and promote action by rele-

vant agencies and local authorities – to help the industry and the rural communities to adapt. The means to this end include 'using the measures available under the European Union's Rural Development Regulation and the Structural Funds' (Government of the National Assembly for Wales 2001: 18).

References

Agri-food Partnership (1999a) *The Welsh Lamb and Beef Sector. A Strategic Action Plan*, Prepared by the Welsh Lamb and Beef Industry Working Group, March, Cardiff: The Welsh Development Agency.

Agri-food Partnership (1999b) *The Welsh Organic Food Sector. A Strategic Action Plan*, Prepared by the Welsh Organic Industry Working Group, March, Cardiff: The Welsh Development Agency.

Agri-food Partnership (2001) *Partnership in Action. Progress Report and Key Priorities for the Wales Agri-food Partnership Towards 2003*, Cardiff: The Welsh Development Agency.

Anderson, K., Eklund, E., Granberg, L. and Marsden, T. (2003) *Rural Development as Policy and Practice*, Helsinki: SSKH Skrifter Research Institute, No 16, Swedish School of Social Sciences.

Department of Environment, Food and Rural Affairs, UK (DEFRA) (2000) *South West European Rural Development Programme: Regional Chapter*, available at http://www.defra.gov.uk/erdp/docs/swchapter.

Government of the National Assembly for Wales (2001) *Farming for the Future. A New Direction for Farming in Wales*, Cardiff: Department of Rural Affairs.

Haskins, C. (2003) *Rural Delivery Review. A Report on the Delivery of Government Policies in Rural England*, available at www.defra.gov.uk/rural/pdfs/ruraldelivery/haskins_full_report.pdf.

Marsden, T. (2003) *The Condition of Rural Sustainability*, Assen: Royal Van Gorcum.

Marsden, T., Lowe, P. and Whatmore, S. (eds) (1992) *Labour and Locality. Uneven Development and the Rural Labour Process*, London: David Fulton.

Marsden, T. and Smith, E. (in press) 'Ecological entrepreneurship: sustainable development in local communities through quality food production and local branding', *Geoforum*.

Murdoch, J. (1992) 'Representing the region: Welsh farmers and the British state', in T.K. Marsden, P. Lowe and S. Whatmore (eds) *Labour and Locality. Uneven Development and the Rural Labour Process*, London: David Fulton.

Murdoch, J., Lowe, P., Ward, N. and Marsden, T. (2003) *The Differentiated Countryside*, London: Routledge.

Parrott, N., Wilson, N. and Murdoch, J. (2002) 'Spatializing quality: regional protection and the alternative geography of food', *European Urban and Regional Studies*, 9, 3: 241–261.

Ploeg, J.D. van der (2003) *The Virtual Farmer: Past, Present and Future in the Dutch Peasantry*, Assen: Royal van Gorcum.

Policy Commission on the Future of Farming (2002) *Farming and Food. A Sustainable Future*, London: Department of Environment, Food and Rural Affairs.

Prodi, R. (2002) 'Foreword', in J. van der Ploeg, A. Long and J. Banks (eds) *Living Countrysides. Rural Development Processes in Europe: The State of the Art*, Lochemdruk: Elsevier.

Smith, E., Marsden, T., Flynn, A. and Percival, A. (in press) 'Regulating food risks:

rebuilding confidence in Europe's food?', *Environment and Planning C: Government and Policy.*

Sonnino, R. and Marsden, T. (in review) 'Alternative food chains and the new agri-food paradigm: towards a new research agenda'.

South West Sustainable Farming and Food Steering Group (2001) *Doing Things Differently. Shaping the Delivery Plan for a Sustainable Farming and Food Industry in South West of England.*

Welsh Development Agency (WDA) (2004) *Agri-food Strategy Review 2004–2007*, draft final report, Cardiff: Wales.

Welsh Office Agriculture Department (WOAD) (1996) *A Food Strategy for Wales*, Cardiff: Wales.

Winter, M. (2003) 'The changing governance of agriculture and food: a regional perspective', Inaugural lecture, 26 March, School of Geography and Archaeology, University of Exeter.

Part II
(De)politicizing practices

5 Reshaping the agri-food system

The role of standards, standard makers and third-party certifiers

Carmen Bain, B. James Deaton and Lawrence Busch

Introduction

Knight and co-workers (2002) define agriculture as the system and processes used in the cultivation of plants, and the raising of animals, for food and other materials. The systems and processes that characterize agriculture are a function of a much broader agri-food system – the set of relationships that coordinates food production by harmonizing the choices made by producers, processors, retailers, food service outlets and consumers. The agri-food system is currently being reshaped by a variety of trends including growing international trade (Kennedy 2000), concerns about food safety (Caswell and Hooker 1996), agriculture's impact on the environment (Ogishi *et al.* 2002), worker health, safety and wages (Gereffi *et al.* 2001), animal welfare (Hobbs *et al.* 2002), and the use of new forms of technology (Lusk *et al.* 2003).

Agri-food *standards*[1] and the corollary set of monitoring and enforcement institutions are emerging as a primary means by which public and private participants (standard makers) influence the character of the agri-food system. For example, the World Trade Organization (WTO) has specified the *Codex Alimentarius* Commission's standards as prima facie evidence in trade disputes concerning competing national health standards (Kennedy 2000). The United States Department of Agriculture has established the National Organic Program to implement standards for the use of 'organic' as a marketing term (USDA 2003). Private food retailers have formed EUREP (Euro-Retailer Produce Working Group) to establish labour, environmental and health standards for the produce they purchase (EUREPGAP 2001). These standards have precipitated an increased role for *third-party entities*. Third-party entities are independent, external institutions that assess, evaluate and certify quality claims based on a certain set of standards and compliance methods (Tanner 2000).

As institutions, standards and their corollary enforcement bodies order relationships among people by defining their rights and their exposure to the rights of others (Schmid 1987). One institutional school of thought, often referred to as Old Institutional Economics (Rutherford 1994), emphasizes that institutions define opportunity sets for participants. The choices and actions available to an individual within these opportunity sets are always conditioned by the opportunity sets of others (Commons 1950; Samuels 1971; Schmid 1987). Within this framework a particular focus is given to identifying distributional issues regarding who participates in decision making and whose interests count (Schmid 1989). This theoretical perspective informs our analysis of the role of standards, standard makers and third-party certifiers in the contemporary agri-food system.

In the first section we describe the WTO and the set of standard-setting bodies upon which it relies to promote harmonized international standards. Here, we emphasize that the ability to participate in both setting and implementing these standards differs between developed and developing nations. The second section discusses the growing use of private standards and third-party certifiers by major retailers to coordinate their supply chains globally. We argue that private standards and certification systems are not scale-neutral for farmers and, in some situations, may lead to smaller farmers losing market share to larger farmers. In the third section we examine the use of microbial standards in the Michigan (US) blueberry industry. We emphasize that food safety standards set by manufacturers may be neither uniform nor science based. In each section we develop the common theme that standards and standard-setting bodies are not neutral or benign means for handling issues of technical compatibility. Instead, standards and standard makers expand inevitably the capacity of some participants and limit the capacity of others to reshape social, political and economic relationships.

Public standards: the WTO

In 1995 the General Agreement on Tariffs and Trade (GATT) Uruguay Round established the WTO, an international organization comprising member governments, to devise rules for international trade. Of particular significance is that the WTO, unlike GATT, has international legal status with enforcement powers similar to the United Nations (UN), and its rules are binding on all 146 members (McMichael 2000). The Sanitary and Phytosanitary (SPS) agreement was established under WTO authority. This agreement establishes SPS measures as well as an SPS committee which comprises member country representatives.

SPS measures provide rules for meeting standards for food safety and animal and plant health while at the same time ensuring that these standards are not overly stringent and, therefore, create a trade barrier (WTO 1998). The principal provisions allow countries to establish their own standard as long as they are based on scientific principles and 'are not maintained without sufficient scientific evidence' (Nestle 2003: 115–116). In addition, the principal provisions encourage countries to 'harmonize' their SPS measures by adopting standards set by several international standard-setting bodies (Kennedy 2000). The most important of these are the *Codex Alimentarius* Commission (*Codex*), administered jointly by the World Health

Organization (WHO) and the Food and Agriculture Organization (FAO), which develops food safety standards; the International Office of Epizootics (OIE), which focuses on regulations for trade in livestock; and the International Plant Protection Convention (IPPC), whose purpose is to prevent the spread of plant pests. The standards set by *Codex*, OIE and IPPC establish the prima facie evidence used by the SPS committee to examine member compliance with the SPS agreement. The standards put forth by *Codex*, OIE and IPPC are public rather than private because, in use, they are imbued with authority through the SPS agreement and the WTO. As a result, the WTO, international standard-setting bodies and the SPS committee constitute a public third-party entity that promulgates (for transacting member countries) agri-food standards, the rules for setting country standards and the basis for dispute resolution.

The particular challenges and concerns faced by WTO member countries in implementing the SPS agreement are likely to differ considerably. Moreover, the intra-country effects of meeting SPS standards will also vary. The asymmetric outcomes of these standards are due in part to the disparity that exists between these nations' agricultural sectors. First, the intensity of concerns expressed by developing countries regarding these standards reflects the fact that food and agricultural products comprise a greater proportion of their exports (Hensen and Loader 2001). Indeed, the share of exports of agricultural products in total merchandise is 18.1 per cent for Latin America, 14.7 per cent for Africa, 10.5 per cent for North America and 9.2 per cent for Western Europe (WTO 2002). Second, growth in fresh and minimally processed products from less developed countries to developed countries has grown rapidly over the past decade (Unnevehr and Roberts 2003). These products tend to fall within the purview of standards governed by the SPS Agreement due to the greater biological risks to the health of humans, animals and plants that these products pose in developed countries (Unnevehr and Roberts 2003). Third, the technical capacity is often limited, and the cost of achieving these standards is high, for most developing countries (Henson and Loader 2001). For example, Henson (Henson and Loader 2001) points to estimates where the Bangladesh frozen shrimp industry spent US$17.6 million between 1997 and 1998 to upgrade plants to satisfy EU and US hygiene requirements. Argentina spent US$82.7 million over the period 1991 to 1996 to achieve disease- and pest-free status so that it could export meat, fruit and vegetables (Hensen and Loader 2001).

Kennedy (2000) argues that developing countries may be relatively disadvantaged by the SPS standard-setting process. Based on his assessment of the EU beef hormone dispute, the Australian salmon dispute and the Japan food quarantine dispute brought before the WTO, Kennedy concludes that the SPS Agreement benefits those who have leverage in setting international standards. This is because 'an international standard presumptively is valid, thereby placing a heavy burden of proof on a complaining WTO member' (Kennedy 2000: 100). Unfortunately, many developing countries 'lack the scientific expertise and resources to influence the debate' and thereby leverage the international standard-setting bodies (Kennedy 2000: 100). This has led some observers (Busch and Bain 2004; Henson and Loader 2001; Kennedy 2000) to argue that for these countries to benefit from the SPS Agreement, they must be able to participate fully in the design, implementation, monitoring and evaluation of standards and its institutions. New harmonized EU standards for aflatoxins in peanuts illustrate that developing countries have a great deal to lose if they are not parties to negotiations regarding standards. The new EU standard, which would reduce health risk by approximately 1.4 deaths per billion per year, could decrease exports in cereals, dried fruits and nuts from African nations by 64 per cent resulting in US$670 million in lost revenues (Otsuki *et al.* 2001).

In some cases, the disadvantages noted by Kennedy may be offset by assistance to developing countries. For example, Article 9 of the SPS Agreement requires that technical assistance be provided to developing countries to help them comply with health and safety standards (Silverglade 2000). Unnevehr and Roberts (2003) highlight several cases where a combination of public and private technical assistance helped a developing country meet the SPS standards in the export market. Exports of Guatemalan raspberries were, for example, re-established in 1999 (after exports had been blocked following an outbreak of *cyclospora*-induced illness in the USA in 1996) with the help of 'technical assistance from the US public and private sectors, in cooperation with a new public-private agency established within Guatemala' (Unnevehr and Roberts 2003: 19). However, Silverglade (2000) argues that, in general, developed countries have been reluctant to live up to their obligations to provide assistance. He concludes that the lack of such support has left developing countries with few choices but to argue in arenas, such as *Codex*, for downward harmonization simply so that they can meet the standards (Silverglade 2000).

Nevertheless, even if technical assistance is forthcoming from developed countries, this would not resolve how the benefits of that assistance should be distributed among various stakeholders. Unnevehr and Roberts (2003) found that technical assistance to meet the SPS standards could influence the restructuring of the exporting industry and lead to a decline in the number of producers who are able to participate in the export market. In the Guatemalan raspberry case noted above, the number of farms that are exporting has dropped to only two (Unnevehr and Roberts 2003), down from eighty-five in 1996 (Calvin *et al.* 2002). The SPS Agreement stipulates that countries are obligated to meet the prima facie standards of *Codex* unless they can demonstrate, using scientific principles, that different standards are necessary to protect human, animal, or plant life or health, and do not function as a trade barrier. Since developing countries have a dearth of technoscientific expertise and financial resources, any attempts at establishing an alternative standard that could potentially mitigate these social and economic disparities is extraordinarily difficult.

Private standards

Private standards are also emerging as a means for coordinating the international agri-food system. Private standards, unlike public standards, make authoritative reference to a firm (or group of firms) rather than a government. In some cases private standards may supersede public ones (Unnevehr and Roberts 2003). For example, fast-food chains, such as McDonald's, Wendy's and Burger King in the USA, have responded to demands by animal rights activists to implement more stringent animal welfare standards, such as more humane animal handling and stunning practices from their meat suppliers (Busch and Bain 2004). In particular, supermarkets appear to be increasingly concerned with setting their own standards for production (Dolan and Humphrey 2000; Levidow and Bijman 2002; Reardon and Berdegue 2002). This growing concern may be due to the increasing concentration of supermarkets and a preference for competing on quality attributes rather than on price (Reardon *et al.* 2001). For example, in 2000, the top five supermarket chains in the USA accounted for over 40 per cent of retail food sales, while in 1993 they accounted for only 20 per cent. During this same period, the market share of the top five chains in France increased from 48 to 61 per cent and in Italy from 11 to 25 per cent (Busch and Bain 2004).

Levidow and Bijman (2002) argue that retailers seek to use their claims for food quality to add and capture market value. They explain that in Europe this has encouraged the development of private standards by retailers regarding farm inputs. For example, retailers set their own criteria which require the reduction of pesticide residues and may even stipulate which pesticides their farm suppliers are allowed to use. Furthermore, in 1998, in lieu of clear EU rules, many of these retail chains initiated labelling for their own branded lines of genetically modified (GM) products. Since then, private standards for GM products have often gone beyond EU requirements. For example, 'in Germany and Austria, the entire industry has moved towards negative labeling, e.g. "GM-free" food' (Levidow and Bijman 2002: 5).

Increasingly, private standard-setting bodies rely on third-party systems of verification to bolster their claims all along the vertical supply chain. Henson and Northen (1998) argue that the rise of retailer own-brand labels encouraged the use of third-party certification schemes. The growth of in-house branding has meant that retailers must accept more responsibility and risk in maintaining food quality (Levidow and Bijman 2002). At the interface between consumers and producers, retailers argue that it is consumers who hold them responsible for the safety of products sold in their stores, particularly in the case of retailer-branded products (USDA/FAS 2001). At the same time, certification schemes allow retailers to impose their own food safety requirements, while passing many of the costs of auditing on to their suppliers (Hensen and Northen 1998). According to Busch and Bain (2004), private certification firms may conduct audits on a range of practices including those related to food safety, food quality, Good Agricultural Practices, Good Manufacturing Practices, and/or Good Handling Processes, labour practices, and/or environmental standards. These third-party certification bodies provide signals in the marketplace about food quality and help 'overcome the potential failures that may emerge in uncertain situations characterized by asymmetric distributions of information' (Deaton 2004). Thus concern at the potential loss of reputation (if a company is found to be using, for example, child labour), and the need to minimize liability (if a food safety outbreak should occur), has not only motivated the development of standards but also certification and labelling schemes that can communicate to customers and consumers the product's quality and safety (Farina and Reardon 2000).

An interesting example of the evolution of private retail standards and third-party certification is that of EUREPGAP. In 1997, several large European retailers began to work together under the EUREP banner to establish a harmonized standard for Good Agricultural Practices (GAP), together with a third-party certification system for the production of fresh fruit and vegetables. EUREP requires all its suppliers to meet quality, safety, environmental and labour standards that are superior to those required by governments. The final standards were presented at the EUREPGAP 2001 Conference, along with a list of accredited certifiers (USDA/FAS 2001). These retailers agreed to work together to develop EUREPGAP as a benchmark standard in order to avoid a situation where suppliers of multiple retailers are required to be certified to multiple standards (USDA/FAS 2001). There are plans to extend EUREPGAP to other products, such as meat and grain. EUREPGAP retailers hope that the standards and certification system will not only ensure that food on their supermarket shelves is safer, but that the initiative will 'reduce the cost of monitoring and certification, by harmonizing dozens of national food safety systems long before legislators can do so under the rubric of the *Codex Alimentarius* or WTO' (Busch and Bain 2004). Since the launch of its standards, the influence of EUREPGAP in global produce supply chains has grown rapidly. To date, more than 13,000 suppliers have been EUREPGAP-certified in thirty-two countries, with many more currently in the process of certification. EUREPGAP is more geographically diffuse than most other certification systems and also has more members. As such, it is considered to be one of the most prominent certification systems in the agri-food sector (Konefal *et al.* forthcoming).

The growth of private food and agricultural standards has asymmetrical consequences for countries and producers. On the one hand, the development of segmented and niche markets due to specific private standards may offer opportunities for producers in certain countries. In particular, opportunities will be available for producers in those countries which can make the appropriate organizational and institutional responses and which have access to the required training and financial resources (Farina and Reardon 2000). For example, since the mid-1990s, increasingly rigorous grades and standards regarding fruit appearance, quality, the environment and packaging demanded by retailers have led to major transformations in the New Zealand apple industry. An illustration of this is the requirement by UK retailers that pip fruit produced for them follow integrated fruit production (IFP)[2] programmes. In response, the New Zealand Apple and Pear Marketing Board (ENZA)[3] began a campaign in 1996 to move away from conventional fruit production and for all export growers to use IFP methods by the 2000 to 2001 growing season (McKenna *et al.* 1999), which they achieved. ENZA recognized that participation in IFP programs was a means to develop quality differentiation for their products in an increasingly competitive world market (McKenna *et al.* 2001). While output is relatively low, their aim of being competitive in quality-differentiated markets has helped New Zealand develop as a world leader in supplying the top end of the retail market in North America, the EU and Japan with premium apple varieties.

On the other hand, since private standards often involve international trade and exports from developing countries (Unnevehr and Roberts 2003), the challenge of meeting private standards is especially pertinent to producers from less developed countries. Dolan and Humphrey (2000: 161) found that success for African producers in the fresh vegetables chain depended on their ability to meet (and exceed) the demands of UK retailers 'for consistency of quality, reliability of supply and due diligence'. They argued that there were few possibilities for exporters to participate in the market if they did not have the investment capabilities to meet these standards (Dolan and Humphrey 2000).

Farina and Reardon's (2000) study of the extended Mercosur countries[4] illustrates how the imposition of stringent, privatized standards had varying intra-country effects on the capacity of different producers to participate in the market. They found that in various agrifood subsectors over the past decade the new standards have driven many small firms and farms out of business, and helped to accelerate industry concentration. Only a minority of producers had the financial capacity to institute the sophisticated and expensive applications of science and technologies, transportation methods and logistics necessary to meet the new standards (Farina and Reardon 2000). In one example, thousands of small dairy operations went out of business in just five years in the extended Mercosur countries. Here, the dairy companies provided some capital and technical assistance to producers to help implement the new standards, but this assistance was targeted at the larger suppliers. In Brazil, this process of consolidation continued as the Brazilian government moved to institute standards as tough as those in the private sector (Reardon and Farina 2002).

The exclusion of particular stakeholders from the negotiations, implementation or evaluation of private standards may accentuate both inter- and intra-country effects by allowing companies to be selective about which standards they wish to follow, with producers and developing countries left to bear the consequences. For example, the desire for particular food quality standards by European consumers and retailers, such as pineapples that are golden and ripen quickly, has led to the overuse of chemical inputs on pineapple farms in Ghana (Blowfield 1999). Consequently, some trade unions and environmental groups have sought greater stakeholder representation in the development and implementation of standards. In this way, they hope to help ensure that standards can better meet the needs not only of shareholders and customers in developed countries but also employees, communities and suppliers in developing countries (Blowfield 1999).

The introduction of private microbial standards in the Michigan blueberry industry

The Michigan blueberry industry provides another example of the growing impact of private standards. The USA and Canada are the world's leading producers of cultivated blueberries, together accounting for nearly 90 per cent of world production, with the State of Michigan producing the largest share (38 per cent) (Michigan Department of Agriculture 2003). In 2000, there were approximately 575 growers in the state, three-quarters of them owning fewer than twenty-nine acres. The industry in Michigan is segmented into two major market categories: approximately 70 per cent of blueberries are for the processed market with the remainder going to the fresh market. Currently, there are around thirty-five growers who run their own processing facilities that clean and pack berries to be sold for further processing by food manufacturing companies, such as Sara Lee. These companies then freeze, can, dry and liquefy the berries for use in hundreds of products ranging from pies and muffins to sauces and entrées. The size and quality of a grower's processing facility varies considerably from small, labour-intensive operations through to the more technologically sophisticated enterprises.

Blueberries in Michigan are governed increasingly by private standards promulgated by food manufacturers and tested by laboratories. Food manufacturers using frozen blueberries have begun to establish their own microbial standards, which upstream blueberry processors are required to meet. Processors are required to send samples to independent laboratories for testing, although there is no uniformity among buyers as to how often to sample (Bain and Busch 2004). The microbial standards are designed to reduce the possibility of outbreaks of food poisoning. However, blueberries in Michigan have not been implicated in any outbreaks of food-related illness. In fact, there are few known cases where pathogens have been linked to blueberries, indicating that, while incidents are possible, they are very rare.[5] The frozen market includes blueberries that are frozen and packaged in polybags for retailers for use in products such as yoghurt and ice-cream. Since the product is not cooked, there is the potential for microbial contamination of the fruit. Hence these buyers have very low or zero-tolerance standards for human pathogens such as *Staphylococcus, Listeria monocytogenes* or *E.coli 0157: H7*.

Private standards set by food manufacturers have emerged alongside traditional institutions for the governing of food safety. Here, standards continue to be set by the United States Department of Agriculture (USDA) and the Food and Drug Administration (FDA). In 1998, the FDA (1998) published guidelines for minimizing microbial hazards on fruits and vegetables, so-called Good Agricultural Practices (GAPs). These GAPs set the standard for addressing microbial food safety hazards common to the growing, harvesting, washing, sorting, packing and transporting of most fruits and vegetables sold to consumers in an unprocessed or minimally processed (raw) form.

The development of private standards for microbial levels highlights a number of distributive issues. The standards promulgated by food manufacturers place the burden of proof and costs on blueberry processors. Moreover, processors find these standards difficult to meet since the specifications vary from buyer to buyer. For example, four different companies have four different limits for the number of allowable colony-forming units (CFUs) of mould, yeast or mesophilic aerobic bacteria per gram. The difference between their standards is considerable; for instance, one company accepts twenty times the number of CFUs per gram for yeasts as does another, competing, company (Bain and Busch 2004).

While the costs of private standards appear to be borne by processors, it is unclear as to which sectors benefit. Widely disparate standards may indicate that there is very little scientific basis to the standards and that the specifications are largely arbitrary.[6] The variation in private standards suggests that a number of supply-side considerations (e.g. costs or position in the vertical supply chain) influence the means by which producers address consumers' demands for safe food. More broadly, the efficacy of end-product testing itself for reducing microbial contamination is contested. One reason is that pathogens are not homogeneous throughout a food; instead they tend to form pockets and are therefore found sporadically. If a pathogen is statistically present in low numbers, it will be difficult to locate. For example, if a pathogen is present at an 0.1 per cent level and some sixty samples of a given lot are examined, there is a 94 per cent chance of not finding the pathogen (Hingley 1998).[7]

The increasing use of private standards to govern the processing of blueberries reflects growing concerns about food safety and liability in the agri-food system. Frozen blueberry manufacturers are in a position to impose private standards on upstream processors. As a result, the burden and cost of meeting standards has been shifted upstream. The benefits of these private standards presumably accrue to frozen food manufacturers (via reduced liability) and downstream consumers who demand food safety. However, as we have pointed out, variability in standards and problems associated with end-product testing suggest uncertain benefits. This highlights two important issues that characterize the emergence of new standards in the agri-food system. First, standards imply certain costs but their benefits are often uncertain. Second, the costs and benefits of standards are likely to be unequally distributed along the vertical supply chain.

Conclusion

Standards and third-party certifiers are part of the institutional infrastructure that coordinates the production and distribution of agricultural products. In this chapter we considered standards in three different settings. First we studied the WTO and its effort to promote standards set by *Codex*, OIE and the IPPC. Second, we discussed how retailers use private standards and third-party certifiers to source quality fresh products from around the globe. Finally, we examined the use of microbial standards in the Michigan blueberry industry. In each case we developed the argument that the burden of standards differs among market participants. Within the WTO, for example, developing countries lack both the financial and technical expertise to participate equally in both the setting and implementation of standards. With regard to retailers, private standards may be imposed that in some situations are not scale-neutral for participating upstream farmers. In these situations, large farmers may gain market share relative to smaller farmers. In Michigan, blueberry manufacturers set standards that affect blueberry processors and these standards appear to be neither uniform nor science-based.

Public discussions regarding standards often portray standards as universally beneficial, citing, for example, their ability to improve food safety, reduce transaction costs or facilitate greater access to markets. Our analysis does not call these benefits into question. Indeed, the normative basis for judging standards is not the subject of this chapter. Instead, we question the 'universal' character of such benefit claims and endeavour to point out, non-pejoratively, that changes in standards (like existing standards themselves) invariably expand the opportunities of some and limit the opportunities of others.

Acknowledgements

This chapter is based partly upon work supported by the National Science Foundation under Grant Number SBR 0094618. Any opinions, findings and conclusions or recommendations expressed in this material are those of the authors and do not necessarily reflect the views of the National Science Foundation.

Notes

1 Standards are documented criteria or specifications, used as rules, guidelines or definitions of characteristics, to ensure consistency and compatibility in materials, products and services. In use, standards become measures by which products, processes and producers are judged.
2 The key production practices that differentiate IFP from Conventional Fruit Production include: 'reduced use of organo-phosphate pesticides, selective and targeted chemical use after monitoring for actional pest levels or disease-presence, and an overall commitment to more "environmentally produced" fruit' (McKenna *et al.* 2001: 158).
3 From 1970 to 1995 ENZA controlled the marketing of the entire export crop. From 1999 ENZA became ENZA Ltd but retained its monopoly status. However, from the 2002 season apple growers had a choice of exporters.
4 Mercosur members include Argentina, Brazil, Paraguay and Uruguay, while Chile is an associate member.
5 Those incidents that have been reported include an outbreak of Hepatitis A in New Zealand in 2002, where blueberries were identified as the source of infection (Jean *et al.* 2003). In addition, a producer was forced to recall an undetermined number of packages of frozen blueberries from California, Illinois and Australia due to contamination with *Listeria monocytogenes* (FDA Enforcement Report 1998).
6 Establishing a standard for *Listeria*, for example, is difficult since it is ubiquitous in the soil and would almost certainly be detected at some level. However, it is unlikely to be of public health significance. Similarly, *staphylococcus* is usually a problem with temperature-abused meat, but not with blueberries (Bain and Busch 2004).
7 The inadequacy of end-product testing for fruit was illustrated in the *Cyclospora* outbreak with raspberries from Guatemala in 1996 when some 850 people were taken ill. All links between particular products and *Cyclospora* were based on epidemiological evidence alone. Methods to test produce for *Cyclospora* are relatively insensitive for detecting low levels of the parasite. The probability of detecting *Cyclospora* with random sampling of all shipments is quite low, since oocysts are generally not evenly distributed in a shipment (Calvin *et al.* 2002).

References

Bain, C. and Busch, L. (2004) *Standards and Strategies in the Michigan Blueberry Industry*, East Lansing: Michigan Agricultural Experiment Station, Research Report 585, March.
Blowfield, M. (1999) 'Ethical trade: a review of developments and issues', *Third World Quarterly*, 20, 4: 753–770.
Busch, L. and Bain, C. (2004) 'New! Improved? The transformation of the global agrifood system', *Rural Sociology*, 69, 3: 321–346.
Calvin, L., Foster, W., Solorzano, L., Mooney, D., Flores, L. and Barrios, V. (2002) 'Response to a food safety problem in produce. A case study of a cyclosporiasis outbreak', in B. Krissoff, M. Bohman and J. Caswell (eds) *Global Food Trade and Consumer Demand for Quality*, New York: Kluwer Academic/Plenum Publishers.
Caswell, J.A. and Hooker, N.H. (1996) 'HACCP as an international trade standard', *American Journal of Agricultural Economics*, 78: 775–779.
Commons, J. R. (1950) *The Economics of Collective Action*, New York: Macmillan.

Deaton, B. J. (2004) 'A theoretical framework for examining the role of third-party certifiers', *Food Control*, 15: 615–619.

Dolan, C. and Humphrey, J. (2000) 'Governance and trade in fresh vegetables: the impact of UK supermarkets on the African horticulture industry', *Journal of Development Studies*, 37, 2: 147–176.

EUREPGAP (2001) *History – 'EUREPGAP Fruits and Vegetables'*, available at http://www.eurep.org/sites/index_e.html (accessed 30 March 2004).

Farina, E. and Reardon, T. (2000) 'Agrifood grades and standards in the extended Mercosur: their role in the changing agrifood system', *American Journal of Agricultural Economics*, 82, 5: 1170–1176.

FDA Enforcement Report (1998) *Recall of Frozen Blueberries*, 30 December.

Gereffi, G., Garcia-Johnson, R. and Sasser, E. (2001) 'The NGO-industrial complex', *Foreign Policy*, July/August: 56–65.

Hensen, S. and Loader, R. (2001) 'Barriers to agricultural exports from developing countries: the role of sanitary and phytosanitary requirements', *World Development*, 29, 1: 85–102.

Hensen, S. and Northen, J. (1998) 'Economic determinants of food safety controls in supply of retailer own-branded products in United Kingdom', *Agribusiness*, 14, 2: 113–126.

Hingley, A. (1998) *Focus on Food Safety. Initiative Calls on Government, Industry, Consumers to Stop Food-Related Illness*, US Food and Drug Administration.

Hobbs, A.L., Hobbs, J.E., Isaac, G.E. and Kerr, W.E. (2002) 'Ethics, domestic food policy and trade law: assessing the EU animal welfare proposal to the WTO', *Food Policy*, 27: 437–545.

Jean, J., Vachon, J.F., Moroni, O., Darveau, A., Kukavica-Ibrulj, I. and Fliss, I. (2003) 'Effectiveness of commercial disinfectants for inactivating hepatitis A virus on agri-food surfaces', *Journal of Food Protection*, 66: 115–119.

Kennedy, K.C. (2000) 'Resolving international sanitary and phytosanitary disputes in the WTO: lessons and future directions', *Food and Drug Law Journal*, 55, 1: 81–104.

Knight, C., Stanley, R. and Jones, L. (2002) *Agriculture in the Food Supply Chain: An Overview*, Gloucestershire: Campden and Chorleywood Food Research Association.

Konefal, J., Mascarenhas, M. and Hatanaka, M. (forthcoming) 'Governance in the global agro-food system: backlighting the role of transnational supermarket chains', *Agriculture and Human Values*, 22, 3.

Levidow, L. and Bijman, J. (2002) 'Farm inputs under pressure from the European food industry', *Food Policy*, 27, 1: 31–45.

Lusk, J. L., Roosen, J. and Fox, J. (2003) 'Demand for beef from cattle administered growth hormones or fed genetically modified corn: a comparison of consumers in France, Germany, the United Kingdom, and the United States', *American Journal of Agricultural Economics*, 85, 1: 16–29.

McKenna, M., Le Heron, R. and Roche, M. (2001) 'Living local, growing global: renegotiating the export production regime in New Zealand's pipfruit sector', *Geoforum*, 32: 157–166.

McKenna, M., Roche, M., Mansvelt, J. and Berg, L. (1999) 'Core issues in New Zealand's apple industry: global-local challenges', *Geography*, 84, 3: 275–280.

McMichael, P. (2000) *Development and Social Change: A Global Perspective*, 2nd edn, Thousand Oaks, CA: Pine Forge Press.

Michigan Department of Agriculture (2003) *Michigan Blueberries*, available at

http://www.michigan.gov/mda/0,1607,7–125–1570_2468_2471_7390–12863
—,00.html (accessed 15 April 2004).

Nestle, M. (2003) *Safe Food: Bacteria, Biotechnology, and Bioterrorism*, Berkeley, CA: University of California Press.

Ogishi, A., Metcalfe, M. and Zilberman, D. (2002) 'Animal waste policy: reforms to improve environmental quality', *CHOICES Magazine* (Autumn): 15–18.

Otsuki, T., Wilson, J. and Sewadeh, M. (2001) 'Saving two in a billion: quantifying the trade effect of European food safety standards on African exports', *Food Policy*, 26, 5: 495–514.

Reardon, T. and Berdegue, J. (2002) 'The rapid rise of supermarkets in Latin America: challenges and opportunities for development', *Development Policy Review*, 20, 4: 371–388.

Reardon, T. and Farina, E. (2002) 'The rise of private food quality and safety standards: illustrations from Brazil', *International Food and Agribusiness Management Review*, 4: 413–421.

Reardon, T., Codron, J.-M., Busch, L., Bingen, J. and Harris, C. (2001) 'Global change in agrifood grades and standards: agribusiness strategic responses in developing countries', *International Food and Agribusiness Management Review*, 2, 3: 421–435.

Rutherford, M. (1994) *Institutions in Economics. The Old and the New Institutionalism*, Cambridge: Cambridge University Press.

Samuels, W. (1971) 'Interrelations between legal and economic processes', *Journal of Law and Economics*, reprinted in W. J. Samuels (1992) *Essays on the Economic Role of Government: Volume 1 – Fundamentals*, New York: New York University Press.

Schmid, A.A. (1987) *Property, Power, and Public Choice. An Inquiry into Law and Economics* (2nd edn), New York: Praeger.

Schmid, A.A. (1989) 'Law and economics: an institutional perspective', in N. Mercuro (ed.) *Law and Economics*, Boston, MA: Kluwer Academic Publishers.

Silverglade, B. (2000) 'The WTO Agreement on sanitary and phytosanitary measures: weakening food safety regulations to facilitate trade', *Food and Drug Law Journal*, 55, 4: 517–524.

Tanner, B. (2000) 'Independent assessment by third-party certification bodies', *Food Control*, 11: 415–417.

Unnevehr, L. J. and Roberts, D. (2003) 'Food safety and quality: regulations, trade, and the WTO', Invited paper presented at the International Conference on Agricultural Policy Reform and the WTO: Where are We Heading, Capri, Italy.

USDA/Agricultural Marketing Services (2003) *The National Organic Program*, available at http: //www.ams.usda.gov/nop/Consumers/brochure.html (accessed 12 November 2003).

USDA/FAS (2001) *European Union Market Development Reports. Horticulture Products Certification 2001*, Report No. E21134: 3.

World Trade Organization (WTO) (1998) *Sanitary and Phytosanitary Measures*, Geneva, Switzerland: World Trade Organization.

World Trade Organization (WTO) (2002) *International Trade Statistics 2002*, Geneva, Switzerland: World Trade Organization.

6 Disciplining the organic commodity

Hugh Campbell and Annie Stuart

Introduction

This chapter examines the politics of regulation and governing that emerge at the intersection between food standards and environmental sustainability. One of the most prominent areas of research increasingly evident in the 'food networks' literature is that relating to food standards. Lawrence Busch and Keiko Tanaka's work on canola, for instance, provides the most frequently cited exemplar of how food standards might be interpreted within the broader Latourian tradition of networks theory. Nevertheless, there are some limitations to this analysis. An intriguing mention of the Foucauldian concept of 'disciplining' commodities is evident, but not explicated, in the work by Busch and Tanaka. This chapter provides the opportunity to examine the 'disciplining' of commodities through a case study of how the organic social movement sought to create a form of governing over organic products.

Another theoretical point of dialogue is shown in the recent work by Richard Le Heron (2003), who provides the opportunity for integrating some of the insights of the regulationist approach within agri-food theory with the food standards literature. Le Heron argues that New Zealand provides a useful case study of the way in which new forms of regulation and governmentality – through what is termed 'audit culture' – are emerging within New Zealand's neo-liberalized agri-food chains. One of his case studies is the New Zealand organics industry as described in Campbell and Liepins (2001).

In this chapter we intend to broaden Le Heron's (2003) engagement with the New Zealand organic industry by revisiting prior research in the light of convergent ideas about food standards, the disciplining of commodities and agri-food governing. We argue that the transition from loose organic social movement to formal internationalized organic production standards demonstrates three theoretically important insights into how agri-food scholars understand specifically *environmental* food standards in neo-liberal space. First, this transition demonstrates the formation of a particular system of governing over organic food production. This system emerged through a

contested, but increasingly elaborate, set of disciplines around organic production. Eventually these disciplines moved out of the control of the local
organic social movement and into the realm of international benchmarking
and audit. Second, as the disciplines increased in their specificity and rigour,
organic production shifted from being a *mutable immobile* – grounded in local
knowledge and praxis – to being an *immutable mobile*[1] using a rather more
abstract set of international benchmarks. Blending Foucault with Latour,
increasing disciplines within organic systems of governing also increased the
immutability of organic knowledges for producers of organic food. Finally,
mobilizing Kloppenburg's discussion of the relationship between mutable
and immutable knowledges and outcomes for sustainable production, evidence is presented of the increasing remoteness of the new disciplined governing system from the actual sustainability demands of local production
environments.[2]

Agri-food theory: food standards, auditing and regulation

Ten to fifteen years ago, agri-food theory emerged as an alternative explanatory framework to that of the new rural sociology. Influenced partially by
the work of Bill Friedland, and also by French regulation theory (together
with the associated insights of Friedmann and McMichael's (1989) food
regimes theory), agri-food theory sought an engagement with commodity
chains, food complexes, filieres, systems of provision and value chains that
linked production of food to processing, distribution and end consumption
(Buttel 1996). It may be argued that the past ten years of agri-food theory
have been characterized by a divergence into two traditions – one retaining
a political economy approach through the use of French regulation theory,
and the other mobilizing a more post-structuralist method through the
food networks tradition (see Goodman and Watts 1997; Marsden and Arce
1995).

This chapter does not aim to revisit the divergence and conflict between
food networks and the political economy/regulationist traditions in agri-
food theory. Rather, it examines an important site of re-convergence between
the two traditions: the analysis of food standards and audit cultures. The
work of Lawrence Busch and co-workers in the food networks tradition, and
Richard Le Heron within a more regulationist political economy tradition,
may be drawn together in parts to allow 're-engagement' between the two
theoretical traditions. These re-engagements demonstrate that the examination of food standards not only addresses a previous lacuna in agri-food
analysis (with credit going to food networks theory for identifying such a
lacuna), but also indicate the way in which sites of regulation around food
standards and audit cultures in agri-food systems might point to important
vanguard forms of governing in neo-liberal food economies (thus retaining
the flavour of regulationist political economy).

Lawrence Busch and Keiko Tanaka's work on the evolution of grades and standards in the rapeseed/canola subsector provides paradigmatic evidence of the importance of food standards to the formation and stabilization of agri-food commodity networks (Busch *et al.* 1994; Busch and Tanaka 1996; Tanaka *et al.* 1999; Tanaka and Busch 2003). Their analysis concentrates on the intense and cross-cutting relationships between agricultural science, industry and policy in the life story of canola (Busch *et al.* 1994; Tanaka and Busch 2003). Over three dozen grades, standards and filters, operating in close and political relationships with science and policy, are applied to construct the commodity called 'canola'. Using Latour's notion of 'technoscience', they create an intricate elaboration of the technical grades and standards that 'discipline' the human and non-human participants in the canola network.

After nearly ten years of work on the canola sector, the body of analysis of Busch and co-workers serves the useful purpose of both justifying its original intent – to highlight the role of human/non-human symmetry exemplified by standards as a key issue for agri-food analysis – as well as raising a number of intriguing issues. The canola research to date hints only at some of the broader politics behind such standards: trade politics and protectionism remain at the margins of their narrative. Similarly, the key non-human actor in food standards literature is the disciplining power of standards themselves. Using a Latourian approach, the food standards literature clearly highlights the power of standards as non-human actors (or actants) in food networks. We argue, however, that if the terrain is shifted into food standards – such as organic standards – where *environmental* goals are intrinsic to the political intent of the standards, other non-human actants in the system become just as important. Standards for organic produce also indicate another realm of Latourian hybridity: the point of food production itself, where biophysical nature and experimentally inclined organic producers can be revealed as temporary co-producers of what constitutes 'organic' food.

Addressing the first of these questions – the potential social and political influences operating in the arena of food standards – takes us back towards the regulationist approach. Richard Le Heron, for instance, makes the case that there are indeed wider implications for the global food economy evident in the evolution of food standards, quality contracts and audit cultures (Le Heron 2003). Le Heron argues that New Zealand, as one of the world's most evolved neo-liberal polities, provides a compelling analytical space for investigating the possible contours of late neo-liberal regulation. Accepting Peck and Tickell's 'second wave' regulationist vision of capitalism under neo-liberalism as being best exemplified as the instability and uncertainty of 'jungle law' (Peck and Tickell 1994), spaces such as New Zealand and Australia may be seen as sites of 'crisis experimentation' in forms of regulation and governing (see also Campbell and Coombes 1999; Campbell and Lawrence 2003).

Le Heron (2003) uses three case studies – organic/IPM systems in horticulture, increasing contractualism in sheep meat supply chains, and the experiment of a Royal Commission on Genetic Modification – as exemplars of two tendencies in New Zealand. First, as some state functions wither, the emergence of agri-food value chains as an analytically important site of action in agricultural relations is confirmed. These value chains form the key site of action for multi-scalar processes ranging from the global (WTO, trade standards, global markets and supply chains) to regional, local and even individually embodied (farmer/growers as co-producers of agri-food systems) scales of action. Accordingly, Le Heron's three case studies may all be seen as experiments (or even vanguard forms) in the vacuum left by neo-liberal decomposition of old forms of governance. Le Heron (2003) suggests that the sum of these parts is a broad trend towards standards, grading, quality contracts and audit culture as a component not only of neo-liberal economic spaces, but also of the individual strategies of agri-food sectors working through such spaces.

Le Heron (2003) therefore provides initial answers to some of the questions about the politics of food standards. These kinds of standards, together with the systems of audit that accompany them, *are* part of a more widespread movement towards a specific regulatory politics in neo-liberal economic spaces. Le Heron (2003) begins to assemble the case that these kinds of strategies relate to the opening up of space (or the closing down of old regulatory forms) characteristic of more neo-liberal spaces in the world economy. However, while Le Heron begins to flesh out the politics around food standards and audit, he only hints at the importance of environmental tensions and contradictions within these new regulatory forms; Guthman (2000) explores such tensions more fully in a critique of the loss of agro-ecological processes in the Californian organic industry.

This chapter proceeds from the theoretical understanding that the analysis of food standards and audit represents something of a convergence between the previously disparate traditions of agri-food theory. Consequently, New Zealand as a site of late neo-liberal experimentation, and the specific case of organic standards (as examined in Campbell and Liepins (2001) and evaluated in Le Heron (2003)), can illuminate this site of theoretical convergence. This case study provides an opportunity to explore Busch, Tanaka and Le Heron's tacitly raised questions: namely, what are the politics that have characterized the emergence of environmental production standards – like organic – and what are the consequences for non-human nature in the emerging (non?) agro-ecology of organic standards?

To undertake this task, we examine the historical development of the idea of organic agriculture in New Zealand; the emergence of the organic commodity; the systems of governing initiated to attempt to discipline the organic commodity; and the eventual shift from mutable immobile (local practice) to immutable mobile (internationally harmonized) standards for organic production.

The emergence of the 'organic intellectuals'

Three sites of evidence/contestation emerged in the discursive combat between organic and Leibig's chemical farming in the early twentieth century. The first took the debate in an anthropological direction, with scientist Sir Alfred Howard's exhaustive studies of agricultural practices in other cultures (especially Asian examples), which showed that composting was essential to maintain long-term soil fertility. The second is exemplified by Lady Eve Balfour, whose Haughley experimental farm set out to demonstrate that artificial fertilizers would ultimately undermine the productivity of England's soils. Finally, debate emerged in the colonies, where British farming practices had been the model for bringing virgin soils into production. While early results were often spectacular, colonial landscapes soon showed evidence of routine over-exploitation, with further stresses and strains obvious as marginal land began to be managed under 'scientific farming' techniques.

New Zealand featured as one prominent site of this debate. The country's founding 'organic intellectual' – Guy Chapman – linked deteriorating health among New Zealanders with poor soil management techniques and a consequent decline in the nutritional value of food. He corresponded with Howard during the 1930s, initiated composting demonstrations, ran a popular radio show and wrote pamphlets. Chapman launched the Humic Compost Club (later renamed the Soil Association) in 1941, some five years before the Soil Association was founded in the UK. His ideas gained strength alongside a wide constellation of discursive concepts, through which prominent writers, educationalists and politicians together defined the more progressive New Zealand that would arise after the Great Depression, with a stronger, independent national identity, better health, an ecological consciousness and more composting.

Stuart and Campbell (forthcoming) provide a review of the ensuing debate between the organic social movement and the emerging nexus of state, government scientists and agri-chemical industries. This early popular manifestation of organic activity in New Zealand has been under-recognized in previous accounts of organic development (see Campbell and Liepins 2001; Campbell and Ritchie 1996). Typically, these prior accounts focus on life after the emergence of an organic food market – which this chapter now marks as the *second* epoch of organic history in New Zealand.

The arrival of the organic commodity

The processes constructing a discourse of organics shifted significantly in the 1970s. During this decade, the Soil Association found itself again part of a wider constellation of compatible activities and groups, this time influenced by the environment movement; European migration to New Zealand; small farming and the settlement of 'lifestyle blocks'; the establishment of cooperatives selling food; and the commune movement (Ritchie and Campbell 1996). This new group of sites circulating the idea of organics shifted the praxis of organic production out of the home garden and into the marketplace. Organic shops, cooperative outlets, folk festivals, organic cafés, vegetarian engagement with the organic and alternative cookbooks all became new locations constructing the idea of organic. Simultaneously, organic discourses expanded beyond their traditional focus on soil health to include novel concerns about pesticides and other chemical residues/additives in food.

Crucially, the discursive field shifted from being that which constructed the idea of *organic* to that which constructed the idea of *organic food*. A differentiation began to open up between sites of organic food *production* (e.g. communes or cooperative urban gardens), and the emerging *'consumer'* of organic food. Effectively, the organic commodity had arrived as a new vehicle to circulate the meaning of organics within the organic discursive field. For organizations and associations participating in the organic movement, controlling or disciplining the organic commodity (and later trying to control the right to define the standards for organic production) was eventually to become the defining feature of their existence.

For many years, however, the market problem for organics – disciplining the commodity – did not present significant difficulties. For all the initial decades of the organic movement's existence, actual tradable organic food existed primarily as an epiphenomenon of the circulation of organic ideas, politics, activism and gardening. The idea of organics circulated in a discursive field made up of identifiable sites – the home gardener, the Soil Association, the alternative social movements involving communes and cooperative gardens, established pamphlets and publications. In short, it was *people* who were characterized as organic first and foremost, and organic people also happened to produce organic food.

Campbell and Liepins (2001) term this people-based system of discipline over the organic marketplace as the 'on trust' system. The disciplining of organic constructed or excluded people, not food, so if consumers and producers were disciplined to the necessary characteristics of the organic social movement, then, automatically, their food was organic as well. The market for organic food would soon start to change this emphasis in significant ways.

Disciplining the commodity

The 'on trust' system began to change towards formal standards for organic production in the 1980s. Both in New Zealand and internationally, the organic movement became concerned about potential appropriation of the organic commodity and the need for some formal system of governance over organic products. The arrival of an organic food market had clearly created the potential for commercial entities to appropriate the term 'organic' and, like all commodities, organic food proved to be all too dangerously flexible and mutable for those wishing to control closely its meaning.

The origins of formal mechanisms of governing may be traced to 1983, when various organic organizations held a combined meeting which established a unified council to represent organic ideas, teach organic techniques, establish standards for organic food and formally certify to those standards (Ritchie and Campbell 1996). Three organizations – the Biodynamic Farming and Gardening Association, the Soil Association and the Doubleday Association – agreed to establish the NZ Biological Producers Council (NZBPC) (later renamed BIO-GRO NZ).

While these activities happened in New Zealand, similar moves were taking place internationally with the establishment of the International Federation of Organic Agriculture Movements (IFOAM). This body attempted to establish exactly the same kinds of disciplines around the term 'organic' that were being attempted in New Zealand. The two scales of activity were directly linked: a leading figure in the NZBPC, for instance, attended IFOAM meetings and was elected to the World Board of IFOAM (later hosting the World Congress of IFOAM in New Zealand in 1994). These organizations initiated a process of regulation and governance for organic food that aimed to discipline both the commodity and its producers, thus hoping to ensure that the main goals of the organic social movement could be maintained. The system of governing that emerged, and enrolled key parties in the organic sector, was based around a new set of disciplines. These were not the old social pressures and self-regulation of the organic social movement, but rather the new texts, prescriptions, standards and formal protocols for organic production, all to be enforced by an impartial inspectorate comprising 'trustworthy' members of the social movement.

The standards for organic food developed in the 1980s differ significantly from those for other products (e.g. canola) in that they were primarily standards for the *production* of organic food. The NZBPC (and IFOAM) considered this focus on production standards essential in order to guarantee both the absence of chemical residues of an undesirable nature, and the presence of good techniques for achieving environmentally enhanced food production. Key to this system of governance, however, is that the production standards are mainly directed towards *controlling inputs* into organic production rather than evaluating the outcomes through direct food testing.[3] This may seem naively trusting, but it suited a system of audit run on limited resources by volunteers. Basically, annual auditing of the practices of growers is cheap, whereas testing food for residues is not.

The development of formal standards was a slow process, and draft standards through the 1980s were vague, comprising a few pages of printed material. Furthermore, as the NZBPC was a non-professional body, meetings to discuss standards and annual inspections of organic properties were conducted by volunteers. The process of developing standards, moreover, was not controlled solely by a central committee in the NZBPC but rather emerged in dialogue between producers and NZBPC inspectors. Growers would quiz inspectors, write to the NZBPC, or discuss techniques at the frequent organic meetings, farm visits and discussion groups. While these were the key human sites of action, it is crucial to this chapter's argument to recognize that what constituted legitimate organic production at this time was also predicated on what was *biophysically possible* in production. Whole sectors of New Zealand agriculture stayed largely outside organic production because biophysical constraints in organic production could not be solved. The kiwifruit sector, for example, was transformed from a moribund laggard to dynamic market leader through the resolution (via a mineral oil) of control of a scale pest in 1991 (Campbell *et al.* 1997). In short, the first standards, being akin to the prior 'on trust' system, was highly *mutable* to pressures and ideas from multiple local parties, as well as being closely aligned with local biophysical conditions for production.

By illustration, within about a decade of the resolution of the scale problem in kiwifruit, high levels of activity in kiwifruit production had produced a large volume of standards and prescriptions, while in contrast, the nascent organic deer sector had been unable to create any dialogue, and hence any standards, through that entire period. In short, local biophysical conditions clearly allowed standards to emerge in dialogue between growers and the NZBPC inspectors and standards' committee. Notable absentees from this process were agencies linked to the New Zealand government, and agricultural scientists. This grassroots 'on the farm' approach made New Zealand strikingly different to the EU, USA or Japan where science and the state played a major role in creating standards.

Over the next ten years, however, a notable transition began to solidify and bureaucratize the standardized and disciplined organic *commodity* at the heart of the organic discursive field. This transition was characterized not only by the emergence of new organic actors, but also by a shift in the terrain where the organic commodity was disciplined. The consequence was a shift from highly mutable (and locally sensitive) organic standards, to much less mutable standards disciplined by international benchmarking and audit.

From organic praxis to organic text

The arrival of corporate actors marked the first significant change for the industry. Campbell (1997) outlines how both Heinz Wattie NZ Ltd and Zespri International Ltd entered the production and distribution of, respectively, organic vegetables and kiwifruit. Within the industry, these new corporate actors converted a significant number of growers to organic production. Organic land area increased even more rapidly as the new exporters targeted larger producers for conversion: all the resulting organic produce was exported, primarily to Japan, the USA and the EU. The overall effect was rapid growth in the industry and increasing grower numbers.

Campbell and Liepins (2001) discuss in detail how organic standards facilitated the arrival of exporters, and led to corporate actors becoming important participants in the definition of such organic standards. In brief, the entry of exporters prompted concern within the NZBPC, which responded by generating more rigorous and detailed textual standards. Governance by prescription emerged as the NZBPC sought to control the perceived negative potential of exporters.[4] Campbell and Liepins (2001) argue that, in reality, shifting the standards from being grounded in local mutable praxis controlled by the organic social movement to being represented in formal text meant that the NZBPC actually opened up the industry for entry by outsiders. Under the new export culture, inspection did not involve the old 'on trust' evaluation of the *organic grower*; rather, the focus shifted to inspection of technical compliance to production standards for *organic food*. Now organic growers were subjected to the disciplines of technical standards, and many long-term adherents abandoned certification to retain their autonomy within the local, uncertified market for organic products (Coombes and Campbell 1998). At the same time, overall numbers of growers increased significantly as corporate firms found that the task of converting new producers had become much easier: growers no longer had to be members of the organic social movement; they just had to be skilled producers who could meet the new disciplines of the organic standard.

As the organic industry began to expand in the mid-1990s, the role of the NZBPC as an umbrella group for the wider organic agriculture movement in New Zealand altered. In 1995, the NZBPC changed its name to BIO-GRO NZ – already the name of the NZBPC organic label. The renamed organization began to use the increased membership levies from corporate exporters to develop a professional inspectorate with remunerated regional inspectors and a full-time manager of certification. By the end of the 1990s, BIO-GRO barely resembled its old incarnation as representative of the organic social movement, and had become primarily a professional certifying agency liaising with company technical officers, professional inspectors and emerging global systems of audit. The 1990s, therefore, were the decade in which the organic commodity became subject to the discipline of what Le Heron (2003) describes as a managerial audit culture.

Effectively, the two-way relationship between organic inspectors and growers in constructing a praxis of sustainable production at the local level was disrupted. As time went by, export growers increasingly became the passive recipients of standards and techniques established in negotiation between inspectors and company officers and shaped by emerging bodies of global market requirements. Second, for growers in the export sector, it was clear that the discursive terrain shaping the content and meaning of 'organic' no longer predominantly involved the actual producers of organic food. Organic production had shifted from being a locally embedded praxis in which the organic grower was the central focus of attention, to 'organic as text': organic production standards that focused attention firmly on the attributes of organic food and disciplined organic growers to the new, more abstract, definitions of what constituted 'good' organic practice.

Global discipline

Since 2000, the Ministry of Agriculture and Forestry (MAF) has added a second layer of audit to the specific processes of the professionalized inspectorate. The MAF has established (at the industry's request) a government audit of certification systems to ensure compliance to international trading standards for organic food. Predictably, this has reinforced key sets of organic standards emanating from export markets such as the EU and USA as the benchmarks for auditing compliance by New Zealand's domestic standards. This international element has created tension in the political relationship between the managerial/audit layer of certification and the local social movement. Increasingly, it appears that globalizing standards from the most powerful blocs in world trade set the primary parameters for disciplining organic production. In response, the local organic social movement, which is still very much present in New Zealand, continues to attempt to exert political influence on organic standard-setting, even as the export-driven audit process slips further from their control. Campbell and Liepins (2001) outline how the local social movement is politically re-engaging with the standards. Many high-profile growers have opted out of the standard-setting process and have set up enclaves of 'on trust' purchasing in local markets. Simultaneously, the Soil Association has lobbied successfully for funding to set up a domestic certification scheme that will enable small and domestically oriented growers to obtain organic certification without incurring the high costs of inspection.

While the shift to more global disciplines in the construction of standards raises many concerns,[5] the final section of this chapter focuses on the key question that Guthman (2000) raises in the Californian context: Has the evolution of this system of governance undermined the ability of organic production to deliver on agro-ecological goals?

Shifting the logic of sustainability in organic production

The original aim of the organic movement was, among other things, to create a more *environmentally sustainable* system of production. This goal was clearly evident during the 1980s, when the NZBPC set about constructing standards for organic production in New Zealand. It endeavoured to create standards that were both compatible with the broader philosophy endorsed by IFOAM, and which also reflected the challenges of sustainable production in the New Zealand context. Forming the standards in light of their mutable quality – responsive to the local – is highly significant (and was reinforced during the ongoing dialogue between inspectors and local growers at the grassroots), specifically because no one expected organic production to achieve immediate 'sustainability' – a final end-state of fully environmentally sustainable production. Instead, organic standards were designed to be benchmarked directly to conventional production. Organic production was defined as being a series of incremental steps beyond conventional production. If the local biophysical constraints made production difficult, organic standards reflected this by setting reasonable (if modest) goals. In contrast, if local production was already environmentally progressive, organic standards set targets to move organic production further towards environmental sustainability. In this model, organic standards therefore only operated in *relation* to other systems of local production.

Behind this system of 'conventional plus' benchmarking was the implicit understanding that organic standards should be shifted continually to make organic production more challenging than conventional production. Each Standards Review process aimed to distil the wisdom of organic specialists as to what constituted reasonable attainments in progress towards greater environmental sustainability in the specific conditions faced by growers in New Zealand. This created a very specific, mutable and locally embedded logic to the way in which organic production would (and could) relate to goals of environmental sustainability. Local knowledge and local relationships between growers, together with the biophysical conditions of crops and growing environments, could closely integrate organic production with local biophysical constraints in a way that growers could actually achieve.

A central problem for new corporate entrants into organic production has been securing supply, and in many sectors high organic standards have been perceived to be a barrier to entry for new organic growers. In those sectors where existing New Zealand standards exceed international benchmarks (especially in livestock and honey production), there is a strong impetus to harmonize the expectations of local standards with the lowest available international common denominator. The central impetus behind standard-setting thus becomes harmonization rather than establishing achievable targets for more sustainable local production. Benchmarking removes the relativity between conventional and organic production at the local level and thereby, in some sectors, brings much larger numbers of conventional growers closer to organic status.

Here Bruno Latour's langauge is apt. Both the food standards literature and some studies on agricultural sustainability (e.g. Kloppenburg 1991) already draw upon Latour's notion of *immutable mobile* knowledges. We contend that the new politics of organic standard-setting has shifted the knowledge structure (and the integration of human and non-human participants) away from its original tight bundling within what Kloppenburg (1991) calls *mutable immobile* knowledge systems. Now, organics operate according to the disciplines of *immutable mobile* sets of international standards. Kloppenburg (1991) argued that when agriculture makes this shift some fundamental relationships in the achievement of agricultural sustainability are disrupted.

While it may be ideal for knowledge, practices and understandings of sustainable production to be generated at the grassroots level, the New Zealand situation does not indicate a complete abdication in favour of global disciplines. Rather, the organic inspector has emerged as a creative political actor attempting to reconcile local conditions with the bureaucratic demands of global discipline.[6]

Conclusion

The New Zealand case demonstrates both the internal dynamics of shifting styles of governing in the organic industry, and the multiple sites of political engagement with this transition. Food standards literature argues that standards operate as non-human actors in food systems, and thus the disciplining effect of standards on all participants should be understood as part of the co-production of food networks. The emergence of the organic commodity in New Zealand, and the perceived need to control this commodity by the organic social movement, has certainly had unexpected outcomes. Organic standards facilitated the entry of corporate actors into the sector, and enabled new actors to engage with – and then exert considerable influence upon – the regulation of organic food production. Clearly, the standards exert disciplines upon growers, exporters, and even consumers – disciplines which some participants have resisted, some have politically engaged with, and from which some have clearly prospered. In these terms, therefore, the New Zealand case supports strongly the central argument of the food standards literature. However, it also allows us to ask some of the bigger questions that seem to reside outside the main focus of the food standards literature.

We argue that organic standards constitute an interesting experiment in governing occurring within the neo-liberalized space of the New Zealand economy. By revisiting the elaboration of new systems of governing in the organic agri-food chain, we have sought both to give weight to Le Heron's wider argument about the regulatory significance of these emergent experiments in agri-food regulation, and to take the argument one step further to engage with issues of biophysical co-production of organic food production. By shifting the disciplining of organic from the mutable immobile realm of co-production between the organic social movement, growers and biophysical nature at the local level, to the wider audit disciplines of immutable mobile global standards, the chances of long-term sustainable outcomes for organic production are clearly diminished – which is becoming a significant concern for some long-term members of the organic social movement. To paraphrase Guthman (2003), the first (and most baroque) of the eco-labels may provide a compelling new regulatory form which exerts considerable power and discipline over many participants in some agri-food chains. However, while it may be gathering its own regulatory momentum to become a quasi-vanguard form in some economic spaces, it still cannot resolve some of the fundamental contradictions of environmental sustainability under neo-liberal market conditions.

Notes

1 The terms are used in the sense in which Kloppenburg (1991) first introduced them to rural sociology. Kloppenburg used Latour's ideas to articulate the broadly held idea within sustainable agriculture circles that specifically *local* knowledge is advantageous for achieving local-level sustainability (see also Hassanein and Kloppenburg 1995). Local knowledge of sustainable farm practice was equated with *mutable immobile* knowledges, while Western agriculture science was characterized as a system of *immutable mobile* knowledge which had no place within itself for the insights of local knowledge.

2 This analysis of New Zealand displays numerous similarities to the processes of standards development in California described by Guthman (1998, 2000) and we use comparisons between the two to inform the following narrative.

3 This is a tendency rather than a cast-iron rule. There is now some testing of organic products post-production in the wealthier export sectors.

4 Regulation by ever more detailed prescription is not a phenomenon that is restricted to organic standards in New Zealand. Guthman (2003: 139) pithily describes organic standards in the USA as 'the most evolved (and therefore baroque) of the eco-labels'.

5 Of which the 1998 USDA attempt to manipulate federal organic standards is but one example (Guthman 2003).

6 A later publication can explore the new political space inhabited by inspectors/auditors in the new Byzantium of globalizing organic standards.

References

Busch, L. and Tanaka, K. (1996) 'Rites of passage: constructing quality in a commodity subsector', *Science, Technology, and Human Values,* 211: 3–27.

Busch, L., Gunter, V., Mentele, T., Tachikawa, M. and Tanaka, K. (1994) 'Socializing nature: technoscience and the transformation of rapeseed into canola', *Crop Science,* 34, 3: 607–614.

Buttel, F. (1996) 'Theoretical issues in global agri-food restructuring', in D. Burch, R. Rickson and G. Lawrence (eds) *Globalization and Agri-food Restructuring: Perspectives from the Australasia Region,* Aldershot: Ashgate.

Campbell, H. (1997) 'Organic food exporting in New Zealand: the emerging relationship between sustainable agriculture, corporate agribusiness and globalising food networks', in B. Kasimis, H. de Haan and M. Redclift (eds) *Sustainable Rural Development,* Aldershot: Ashgate.

Campbell, H. and Coombes, B. (1999) 'Green protectionism and the exporting of organic fresh fruit and vegetables from New Zealand: crisis experiments in the breakdown of Fordist trade and agricultural policies', *Rural Sociology,* 64, 2: 302–319.

Campbell, H. and Lawrence, G. (2003) 'Assessing the neo-liberal experiment in antipodean agriculture', in R. Almås and G. Lawrence (eds) *Globalization, Localization and Sustainable Livelihoods,* Aldershot: Ashgate.

Campbell, H. and Liepins, R. (2001) 'Naming organics: understanding organic standards in New Zealand as a discursive field', *Sociologia Ruralis,* 41, 1: 21–39.

Campbell, H. and Ritchie, M. (1996) 'History of the New Zealand organic movement', in H. Campbell (ed.) *Recent Developments in Organic Food Production in New Zealand: Part 1, Organic Food Exporting in Canterbury*, Dunedin: Department of Anthropology, Otago University.

Campbell, H., Fairweather, J. and Steven, D. (1997) *Recent Developments in Organic Food Production in New Zealand: Part 2, Kiwifruit in the Bay of Plenty, Studies in Rural Sustainability No. 2*, Dunedin: Department of Anthropology, Otago University.

Coombes, B. and Campbell, H. (1998) 'Dependent reproduction of alternative modes of agriculture: organic farming in New Zealand', *Sociologia Ruralis*, 38, 2: 127–145.

Friedmann, H. and McMichael, P. (1989) 'Agriculture and the state system: the rise and decline of national agricultures, 1870 to the present', *Sociologia Ruralis*, 29, 2: 93–117.

Goodman, M. and Watts, M. (eds) (1997) *Globalising Food: Agrarian Questions and Global Restructuring*, London: Routledge.

Guthman, J. (1998) 'Regulating meaning, appropriating nature: the codification of Californian organic agriculture', *Antipode*, 30, 2: 135–154.

Guthman, J. (2000) 'Raising organic: an agro-ecological assessment of grower practices in California', *Agriculture and Human Values*, 17: 257–266.

Guthman, J. (2003) 'Eating risk: the politics of labeling genetically engineered food', in R. Schurmann and D. Doyle (eds) *Engineering Trouble: Biotechnology and Its Discontents*, Berkeley: University of California Press.

Hassanein, N. and Kloppenburg, J. (1995) 'Where the grass grows again; knowledge exchange in the sustainable agriculture movement', *Rural Sociology*, 60, 4: 721–740.

Kloppenburg, J. (1991) 'Social theory and the de/reconstruction of agricultural science: local knowledge for an alternative agriculture', *Rural Sociology*, 56, 4: 519–548.

Le Heron, R. (2003) 'Creating food futures: reflections on food governance issues in New Zealand's agri-food sector', *Journal of Rural Studies*, 19: 111–125.

Marsden, T. and Arce, A. (1995) 'Constructing quality: emerging food networks in the rural transition', *Environment and Planning A*, 27: 1261–1279.

Peck, J. A. and Tickell, A. (1994) 'Jungle law breaks out: neoliberalism and global–local disorder', *Area*, 26, 4: 317–326.

Ritchie, M. and Campbell, H. (1996) 'Historical development of the organic agriculture movement in New Zealand', in H. Campbell (ed.) *Recent Developments in Organic Food Production in New Zealand: Part 1, Organic Food Exporting in Canterbury, Studies in Rural Sustainability, Research Report No. 1*, Dunedin: Anthropology Department, University of Otago.

Stuart, A. and Campbell, H. (forthcoming) '"Business as usual": contextualising the GM/organic conflict within the history of New Zealand agriculture', *New Zealand Sociology*.

Tanaka, K., Juska, A. and Busch, L. (1999) 'Globalization of agricultural production and research: the case of the rapeseed subsector', *Sociologia Ruralis*, 39, 1: 54–77.

Tanaka, K. and Busch, L. (2003) 'Standardization as a means for globalizing a commodity: the case of rapeseed in China', *Rural Sociology*, 68, 1: 25–45.

7 Governing conflicts over sustainability

Agricultural biotechnology in Europe

Les Levidow

Introduction

Both 'governance' and 'sustainable development' have become key terms in policy debates. These terms have particular salience to techno-scientific controversies in Europe, where protest has challenged the legitimacy of regulatory procedures and innovation priorities. In the case of genetically modified (GM) crops, for example, critics have counterposed 'sustainable agriculture', while agbiotech companies have appropriated the same term to promote their own products. As this chapter will argue, divergent views of sustainability underlie the conflicts over biotechnological innovation and regulation. Governments have extended regulatory procedures and public consultation, sometimes in the name of 'governance', which denotes broader forms of conflict management. I draw upon a case study of European efforts to govern conflicts over GM crops as a sustainability issue. A focus on the late 1990s provides a snapshot of longer term policy changes still underway.

Analytical concepts

Prior to the case study, it is necessary to examine the two policy terms – 'sustainable development' and 'governance' – as analytical concepts.

Sustainable development

'Sustainable development' has become a central concept for public debate and government policy. Since the term was popularized by the Brundtland Report (1987), its meanings have become more diverse and contested. Sustainable development has been widely promoted as a means to achieve environmental sustainability (Dobson 1996). Often the environmental aspect has been distinguished from social and economic sustainability, yet such distinctions can be misleading because all three aspects are linked within any view of sustainability. Fundamentally at issue is how resources should be conceptualized, valued, managed, preserved or consumed – to sustain what kind of society, economy and environment?

Social science has devised various ways to classify views of sustainability. A relational model is necessary for policy analysis – that is, for analysing how various political forces seek allies, undermine opponents and thus attempt to influence policy. For that analytical purpose, Woodhouse (2000) classified divergent views of sustainability as a three-part taxonomy – neoliberal, people-centred, and an environmental management which mediates conflicts between the other two.

In brief, the three views may be summarized as follows (see Table 7.1):

1 Neoliberal (or market-driven): develop eco-efficient technologies and 'green' products to exploit natural capital in ways compatible with the market system, thus enhancing economic competitiveness and environmental protection at the same time (e.g. Schmidheiny 1992).

2 People-centred (or community): devise rules to protect common goods from over-exploitation, as a basis for communities to link producers with consumers, thus resisting industrialization and economic integration into global commerce (e.g. Sachs 2003).

3 Environmental management: enhance the carrying capacity of future ecosystems through technological advance, social reorganization, negotiated rules for resource usage, performance standards and so on; regulate cultivation methods so that they do not undermine agricultural resources (e.g. Brundtland 1987; CEC 2001).

From the above taxonomic perspective, it may be asked: In the case of GM crops, how do the conflicts relate to divergent views of sustainability? What changes occur in regulatory criteria? And why?

Governance

In the political science literature, governance is often understood as co-operative means to deal with common problems and conflicts. For example, governance involves social institutions 'capable of resolving conflicts, facilitating cooperation, or, more generally, alleviating collective-action problems in a world of interdependent actors' (Young 1994: 15). Similarly, governance has been described as 'a continuing process through which conflicting or diverse interests may be accommodated and co-operative action may be taken' (CGG 1996: 2).

Why has the term 'governance' become so prominent in the past decade or so? Although it can simply describe efforts at broader inclusion or participation, the term has more specific origins and meanings. Often governments have invoked international legal and economic imperatives such as 'free trade', especially to override national procedures and sovereignty. As a classic example, global trade rules have been designed to promote regulatory harmonization for trade liberalization. The consequent rules 'effectively narrow the menu of regulatory choices open to governments' (Newell 2003:

Table 7.1 Divergent views of sustainable development

View	Neoliberal (or marketization)	Environmental management	Community (or people-centred)
Led by	multinational companies	government agencies	small-scale producers
Problem definition	inefficiency, depletion of environmental capital	environment/development falsely separated; global interactions	undemocratic institutions; profit-driven innovation
Concept of nature	capital to be invested; assets providing environmental services	eco-support system, human habitat	harmonious balance and/or commons to be shared
Sustain what?	natural capital, substitutable by human capital	optimum resource usage	communities as guardians and beneficiaries of commons
Economic aims	compete better in market for green commodities	economic growth through socio-technical reorganization to increase carrying capacity	enhance livelihoods of small-scale producers
Solution	eco-efficiency to reduce pollution and reap cornucopia	negotiated rules and standards; international cooperation	link producers with consumers
Expertise	R&D for clean products	interdisciplinary networks to model and predict environmental effects	know and work with nature; use local resources

61, 64). More generally, constraints on government are reproduced through 'a discourse of technical-rational knowledge' – that is, by representing all problems as amenable to technical solutions (Ford 2003: 124–125).

Such rhetorical-technocratic imperatives have often backfired, especially by provoking strong protest. This has led governments or international bodies to develop more participatory forms of governing, such as strategies to incorporate dissent. Global governance 'can be seen as a product of two phenomena: the pursuit of neoliberal forms of globalization, and the resistance to such centralization of power' (Paterson *et al.* 2003: 2). From those perspectives we may ask: In the case of GM crops, how does neoliberal globalization generate legitimacy problems and thus efforts to solve these through processes of governing? How do such efforts define collective-action problems?

GM crops as contested sustainability

GM crops have intersected with a wider debate over how to remedy problems which result from intensive agricultural methods. Since the 1980s, biotechnology companies have portrayed their GM crops as environmentally friendly products. Exemplifying a neoliberal view, proponents emphasize that GM crops offer eco-efficiency benefits – by minimizing agrochemical usage, deploying resources more efficiently, increasing productivity, and so enhancing economic competitiveness. This scenario presumes a homogeneous agri-environment as an economic resource for industrial production.

Industry R&D programmes diagnose inefficient agricultural inputs as the problem, which can be solved by precise genetic changes in crops. These link economic competitiveness and environmental efficiency. From those perspectives, society faces the risk of forgoing the crucial benefits that biotechnology can bring. Such arguments exemplify wider links between economic globalization and technological determinism (Barben 1998: 417).

In contrast, critics' arguments have exemplified community views, for example, by defending the agri-environment as common resources and farmers' skills in using them. They have argued that GM crops impose unknown ecological risks, reduce the biodiversity of plant cultivars, subordinate R&D to commercial criteria, generate selection pressure for resistant pests, and promote the further industrialization of agriculture (e.g. Haerlin 1990). They warn against a 'genetic treadmill' by analogy to the agrochemical treadmill – whereby pests develop resistance to pesticides, companies try to develop alternatives faster than the resistance, and farmers become more dependent upon chemical solutions. Moreover, some critics diagnose the problem as intensive monocultural practices which attract pests and disease, while eliminating plant and insect biodiversity which could otherwise help to protect crops.

By the late 1990s, partly in response to critics, the biotechnology industry recast sustainability in its own image of intensive monoculture. For example, inefficient inputs were cited to explain the problems of food insecurity and consequent environmental degradation in poor countries. As a remedy, GM crops would help to increase agricultural productivity, thus increasing production and/or decreasing land requirements and degradation. Other arguments have been more relevant to industrialized countries; GM crops have been portrayed as complementary to Integrated Pest Management (IPM), or even as IPM in themselves (Levidow *et al.* 2002).

Using the slogan 'Creating value through sustainability', the Monsanto Company links market competition, use values, environmental protection and food security. According to its *Report on Sustainable Development*: 'The problem is often framed as a choice: either feed a rapidly growing population ... or preserve natural habitats for biodiversity. But we can do both by continuing the progress of high-yield agriculture' (Monsanto 1997: 16). According to Monsanto, GM crops substitute intelligence for energy and materials: 'Our products create value for our customers by helping them to combine profitability with environmental stewardship. For product impact, this means: more productive agriculture, more soil conservation, less insecticide use, less energy, better habitat protection' (ibid.). In particular, 'inbuilt genetic information' helps GM crops to protect themselves from pests and disease. Herbicide-tolerant crops facilitate no-till agriculture, which 'decreases soil erosion, nutrient and pesticide runoff, as compared to conventional tillage' (Magretta 1997).

According to Novartis, GM insecticidal maize 'contributes to sustainable agriculture through savings on mineral fertilisers, fossil fuels and pesticides' (Novartis 1998). Such arguments exemplify the company's general perspective on intensifying agriculture in more benign ways:

> Sustainable intensification of agriculture can be defined as follows: The use of practices and systems which maintain and enhance: a sufficient and affordable supply of high quality food and fibre, the economic viability and productivity of agriculture, the natural resource base of agriculture and its environment, and the ability of people and communities to provide for their well-being.
>
> (Imhof 1998)

Here the term 'community' is appropriated as an agent of eco-efficient intensification.

Likewise, the term 'biodiversity' has been recast in the image of GM crops. Biotechnology bears 'the prospect of an artificially created biodiversity' in several ways; it seeks to 'smooth out' nature as the means to attain a genetic-level control (Krimsky and Wrubel 1996). Thus, genetic modification changes the terms of reference for what counts as diversity along neoliberal lines of marketizing nature. According to proponents, GM crops provide a greater variety of genetic combinations, which thereby increase biodiversity – redefined as laboratory simulations of natural properties.

Conflicts emerge over GM crops

In Europe, 'sustainable agriculture' has been framed by distinct cultural values, linking the quality of food products, rural space and livelihoods. Although chemical-intensive methods prevail in Europe, the countryside there is regarded increasingly as an environmental issue, variously understood (e.g. as an aesthetic landscape, a wildlife habitat, local heritage, a stewardship role for farmers, and their economic independence). These values conflict with neoliberal models of agriculture as a contest for greater productivity and economic competitiveness.

In European national debates over GM crops, 'risk' discourses have been central, though often linked with 'sustainable agriculture'. Until the mid-1990s Europe had little such debate over GM crops, except in Germany and Denmark. Later, intense conflicts emerged in some other countries. Protest was driven mainly by activists from environmentalist and farmer groups; these catalysed broader opposition networks, as well as scientists' networks which raised doubts about safety claims. This section surveys national features of the Europe-wide debate and protest, illustrating various concepts of sustainability. The subsequent section will analyse regulatory responses during the same period.

Protest emerges

Since the 1980s German NGOs have largely opposed biotechnology. They highlighted its reductionist model which diagnoses social problems as genetic deficiencies. They criticized a 'technology-induced' approach, which simply evaluated risks and benefits of GM crops. NGOs counterposed a 'problem-induced' approach, which would compare such products to other potential weed-control methods as alternative solutions to agricultural problems. But this proposal was marginalized (Gill 1993). NGOs also voiced their concerns in public hearings but were largely dismissed as irrational by officials (Gill 1996).

Germany's policy has been driven by a neoliberal framing of biotechnology as a *Hoffnungsträger* (hope-carrier) – that is, as an essential tool for R&D investment, innovation, a stable job market and international competitiveness. Protesters have emphasized that GM crops threaten 'nature' – popularly associated with forests in Germany, though linked little to agriculture. Such polarization continued through the 1990s (Dreyer and Gill 2000). Neoliberal policy assumptions were finally opened up for debate in 2001, when the Red–Green coalition government initiated the *Diskurs grüne gentechnik*, high-profile public discussions about how agbiotech may relate to sustainable agriculture.

In addition, since the mid-1980s in Denmark, many NGOs questioned whether GM herbicide-tolerant crops would be a step towards sustainable agriculture. They obtained funds to organize an educational campaign to stimulate a national debate, linked with a Consensus Conference on agricultural biotechnology. Trade unions generated further debate on advantages and disadvantages. They distributed material which posed questions about sustainable agriculture: for example, would GM crops alleviate or aggravate the existing problems of crop monocultures (Elert 1991: 12)? In response to that early debate, Denmark's 1986 biotechnology law nearly banned the environmental releases of GMOs, while affirming the general aim of 'sustainable development', like all environmental legislation in that period.

Since the 1980s, Denmark has had a policy to reduce agrochemical usage, especially so that ground water may be used safely as drinking-water. Citing that policy aim, NGOs have demanded risk assessments which evaluate the long-term implications of GM crops for herbicide usage and residues. They successfully pressed the Danish Parliament to raise such questions about herbicide-tolerant crops. In response, the Environment Ministry adopted broad risk-assessment criteria along those lines (Toft 2000). The Danish approach valued ground water as a common resource, implicitly linked with more extensive cultivation methods which use fewer pesticides. Thus environmental management somewhat accommodated a community-type view of public goods.

In Europe GM crops reached the commercial stage amid a wider debate over the future of agriculture. The 1996 'mad cow' crisis undermined the credibility of safety claims for food products. It also aggravated a prior suspicion towards 'factory farming'. This phrase originally denoted agribusiness production-line approaches to animal husbandry, including the caging and 'feedlotting' of animals; it was later extended to intensive methods in general, even for crops.

Anti-biotechnology activists throughout Europe catalysed a wide-ranging risk debate about the intensive methods prevalent in the agro-food chain. Environmental NGOs emphasized unpredictable risks as grounds for a moratorium on commercial use of GM crops (e.g. FoEE 1996–1998). Environmental issues were also taken up by consumer NGOs. Protest linked GM food with environmental risks of cultivating GM crops. Many people boycotted GM food as a way to 'vote' against agricultural biotechnology, in lieu of a clear democratic procedure for a societal decision about a contentious technology. By the late 1990s, in response to consumer and environmentalist protest, most major European retail chains had excluded GM ingredients from their own-brand products (Levidow and Bijman 2002).

By the late 1990s GM crops were being debated as to whether their associated agricultural methods complement or contradict 'sustainable agriculture' – a term that now had diverse meanings (for example, eco-efficiency, Integrated Crop Management, organic farming, peasant autonomy). Eco-efficiency arguments were often cited to promote GM crops.

Such benefits were proclaimed at a time when commercial use had hardly begun in Europe. According to an EU committee, biotechnological solutions are 'guaranteeing yields, helping to cut the use of plant health products in combating pests and diseases, and creating quality products'. Such efficiency extends even to regulatory science: thanks to its precise techniques, genetic engineering 'allows more accurately targeted risk prediction', argued the committee (EcoSoc 1998).

National debates over agbiotech

From an eco-efficiency standpoint, expert evaluation could readily endorse GM crops. In Spain, which had little protest, its national advisory committee implicitly considered their benefits for environmental sustainability. Benefits were defined as any improvement over present practices – for example, the potential for herbicide-tolerant crops to reduce herbicide usage, and likewise for Bt insecticidal crops to reduce insecticide usage (Todt and Lujan 2000). In other countries, however, the evaluation was more stringent or negative.

In Austria GM crops symbolized a threat to organic agriculture and thus to national values. Even before GM crops became an issue there, the Austrian government was promoting organic farming – as ecologically sound, as 'quality' products, and as an economically feasible market-niche alternative for an endangered national agriculture. This anti-biotech scenario of 'competitiveness' contrasted with the pro-biotech imperative to increase agricultural productivity. Some officials regarded agricultural biotechnology as a threat to the environment and an obstacle to sustainability. Austrian regulators compared potential environmental effects of GM crops unfavourably to methods which use no agrochemicals (Torgerson and Seifert 2000). Austria is among several countries or regions which have promoted 'GMO-free zones' as a means to protect the heritage and biodiversity of European agriculture (cited in FoEE 2000).

In the UK anti-agbiotech critics drew an analogy between GM crops, industrialized agriculture and the market pressures which led to the BSE crisis. Critics warned that broad-spectrum herbicides, for which herbicide-tolerant GM crops are designed, could harm wildlife habitats near agricultural fields. On these grounds, the government's own nature conservation advisers had demanded a delay in commercial use. The Consumers Association attacked the agro-food industry for its 'unshakeable belief in whizz-bang techniques to conjure up the impossible – food that is safe and nutritious but also cheap enough to beat the global competition' (McKechnie 1999).

UK farmers were divided or ambivalent. The National Farmers' Union initially supported GM crops as an important tool for economic competitiveness, but later it became more cautious. Early dissent came from a split-off called the Small and Family Farm Association. In 1998 the Soil Association declared that crops must have no GM 'contamination' in order to be certified as organic, and this became an EU-wide standard.

In opposing GM crops, some critics counterposed less intensive methods

– as a future alternative scenario, and as a baseline for judging the environmental effects of GM crops. According to UK environmental consultants, for example, these products became a focus of public pressure because they are designed for an 'increasingly intensive monoculture'. Therefore, GM crops should be evaluated in a wider debate about sustainable agriculture, 'not just relative to today's substantially less-than-sustainable norm' (Everard and Ray 1999: 6).

In France in the mid-1990s, anti-GM activists catalysed a national debate. They launched a scientists' petition, which emphasized unknown risks and advocated a moratorium on GM crops; many prominent scientists signed the petition. Some critics focused on GM herbicide-tolerant oilseed rape, which could readily generate herbicide-tolerant weeds and thus complicate the use of herbicides. Innovation research on such products was abandoned by the Institut National de la Recherche Agronomique (INRA).

In the late 1990s the French debate soon expanded from 'risk' to sustainability issues. Some industrial-type farmers initially sought access to GM crops as a means to enhance their economic competitiveness. Others, affiliated to the Coordination Paysanne Européenne, regarded such products as a threat to their skills and livelihoods. According to French peasants' leaders, GM crops pose risks to their economic independence, to high-quality French products, to consumer choice and even to democracy. This vision resonated with the trend towards producing French food as *produits de terroir*, a label which denotes its origin from specific localities and peasant cultivators.

When peasant activists were prosecuted for sabotaging stores of GM grain, they used the trial to gain public support for their attack on industrialized agriculture. As an alternative future, they argued, 'Today, more and more farmers lay claim to a farmer's agriculture, which is more autonomous, economic, and which integrates problems associated with the environment, employment, and regional planning' (Bové 1998). Against the commoditized inputs of multinational companies, they counterposed their own *paysan savoir-faire* (Heller 2002).

As in France, Italian anti-GM critics sought to protect the agro-food chain as an environment for specialty products. The Italian Parliament had already allocated subsidies to promote local crop varieties, *prodotti tipici*, and now foresaw these being displaced by GM crops. According to a parliamentary report, the government must 'prevent Italian agriculture from becoming dependent on multinational companies due to the introduction of genetically manipulated seeds'. Moreover, when local administrations apply EU legislation on sustainable agriculture, they should link these criteria with a requirement to use only non-GM materials (Camera dei Deputati 1997, cited in Terragni and Recchia 1999).

In the Italian Parliament and government, anti-biotechnology arguments were led by members of the Green Party, which headed the Environment Ministry after the Olive Tree Coalition won the 1997 election. These bodies adopted arguments from Coltivatori Diretti, a million-strong union of mainly small-scale farmers. Its members regarded GM crops as threats to local specialty food products and to crop biodiversity (Terragni and Recchia 1999).

Thus divergent cultural understandings underlay the controversy over GM crops. In various ways around Europe, claims for environmental safety or benefits rested on an eco-efficiency account of sustainability – for example, reductions in pesticide usage. This conflicted with other accounts, emphasizing farmer independence, producer–consumer relationships, land-use patterns and so on.

Alternatives stimulated

As an alternative to industrialized methods, agricultural extensification originated in concepts of 'harmonious control', later 'integrated control', and eventually 'Integrated Pest Management' (IPM). Along with a shift towards biological crop-protection agents, this also meant changes in agronomic practices and farm structure. All these changes draw upon and stimulate research into 'agro-ecology', especially in Europe (e.g. Greens/EFA 2001).

Public protest has given further stimulus to such alternative methods. Food retail chains require and help farmers to adopt cultivation methods which avoid pest problems and so reduce the need for agrochemicals. They promote IPM, which enhances knowledge of how best to use various methods and inputs (EUREP 1999).

Through some IPM methods, farmers could gain independence from purchased inputs from suppliers. Such efforts diverge from intensive agricultural models. Retail chains fund research on soil-management methods which strengthen plant resistance to pests and disease. Organic food lines are expanded by supermarket chains; organic breeding institutes develop pest-tolerant seeds which may be more durable in the face of novel pests (Levidow and Bijman 2002).

The agro-food industry has undergone pressure to change not only the characteristics of products, but also the concept of innovation. Beyond product-based solutions, different cultivation processes are developed. By 2001 some governments were giving more financial support for research on such alternatives. Consequently, future scenarios for European agriculture are not limited to conventional versus GM inputs. Both options are challenged by a debate over what kind of agriculture and society is required. As environmentally less harmful methods are developed for crop protection, these alternatives serve as more stringent comparators than the chemical-intensive methods which underlay early safety claims for GM crops.

Regulatory procedures as conflict mediation

The EU had approved some GM crops for commercial cultivation in the mid-1990s, when safety claims rested on a neoliberal view of sustainability.

In the late 1990s public protest led member states and the EU overall to reopen the original basis. Mediating the conflict, regulatory procedures moved towards more stringent criteria, which potentially favoured comparisons to less intensive cultivation methods (for detailed references, see Levidow and Carr 2000).

Safety approval disputed

For regulating GMOs, EC legislation sought to link environmental protection with market integration by overcoming internal trade barriers. As rationales for Community-wide legislation, proponents cited the prospect that diverse national rules could impede the internal market or that GMOs could cross national boundaries. To address those problems, the Deliberate Release Directive aimed to 'establish harmonized procedures and criteria' for assessing GMO releases, so that any product approval would apply throughout the European Community. Member states had a duty to ensure that GMOs did not cause 'adverse effects to human health or the environment' (EEC 1990: 15). However, the practical definition of 'adverse effects' later proved to be contentious and thus difficult for achieving harmonized criteria.

In the mid-1990s the EU regulatory procedure came under political pressure to approve GM products. Industry-wide lobby groups warned government that companies would shift R&D investment to North America if product approvals were unduly delayed. The EC Directive itself came under attack for stigmatizing GMOs, thus disadvantaging 'European' biotechnology and its competitiveness.

At European and national levels, governments promoted biotechnology on several grounds. According to officials such technological development would attract R&D investment, enhance the efficiency of European agriculture and reduce the environmental impacts of agriculture. Economic arguments came especially from the UK and German governments. Politicians warned against the potential loss of economic and environmental benefits from GM crops. 'Completing the internal market' was sometimes linked with 'free trade' agendas and proposals to liberalize European agriculture.

Within that neoliberal policy framework in the mid-1990s, many national regulators accepted safety claims by companies while acknowledging that GM crops could cause some undesirable effects. If weeds acquired tolerance to herbicides, or if insects acquired resistance to GM toxins, then such inadvertent effects would undermine the efficacy of the corresponding control agent. These 'genetic treadmill' scenarios were conveniently classified as 'agricultural problems' rather than as environmental harm; moreover, other pest-control methods would still be available. Current options were regarded as interchangeable and therefore dispensable, regardless of whether they might be deemed environmentally preferable.

By defining harm in narrow ways, safety claims could treat the European agri-environment as a homogeneous resource for intensive monoculture, by analogy to the US model. GM crops were judged to cause no more harm than the most agrochemical-intensive cultivation methods. In addition, there was no government responsibility for evaluating the effects of changed herbicide practices, for example, a switch from selective to broad-spectrum herbicides. On that basis, EU-wide approval was granted to a GM herbicide-tolerant oilseed rape and insect-protected maize in 1996 to 1997. Dissent came from several EU member states – particularly Denmark, Austria and Sweden. They demanded that the risk assessment should consider a broader range of plausible effects. Some countries also emphasized the overall environmental implications of spraying broad-spectrum herbicides on the crop.

Responses to national protest

After the first shipments of GM soya reached Europe in late 1996, public protest erupted against GM crops, especially in the UK and France. Protestors associated agbiotech with an ominous 'globalization', including greater control by multinational companies. Pressures to industrialize agriculture were associated with the 1996 'mad cow' crisis. Earlier safety assumptions were challenged, and national objections gained strength. In 1998 the EU Environment Council decided that henceforth risk assessments must include any 'indirect effects' of changes in agricultural management. This accommodated UK demands to evaluate effects of herbicide-usage patterns on farmland biodiversity.

In addition, the prospect of a genetic treadmill, formerly marginalized as an 'agricultural problem', was now treated as a risk to be managed and prevented. This policy was implicit in the 1998 EU approval of an insecticidal crop, and was explicit in decisions by France and Spain to require monitoring. In such ways, governments and industry devised further controls on GM crops. These included measures to limit the spread of herbicide-tolerance genes, to limit insect resistance, and to monitor herbicide-tolerant crops for potential harm from broad-spectrum herbicides.

The UK funded measures for testing the overall effects of herbicide usage on biodiversity near fields. 'Farm-scale evaluations' were designed to simulate the practices of commercial farmers, to compare GM herbicide-tolerant crops with previous practices, and so to gain more evidence about broad-spectrum herbicides. Representing views of various environmental groups, nature conservancy agencies were incorporated into the scientific steering committee. These agencies proposed that the experimental design should include non-GM fields which use relatively less intensive farming methods, to provide a more stringent baseline for evaluating the effects of spraying GM crops. The ultimate design incorporated their proposal.

Broader bodies were established to discuss regulatory criteria as policy issues. The UK established an Agriculture and Environment Biotechnology Commission to provide advice on strategic issues, such as definitions of environmental harm and criteria for sustainable agriculture. Likewise, France established a new body to advise the Environment Ministry on general issues, as well as a *biovigilance* committee to evaluate the methods for environmental monitoring of GM crops. Also in France the Parliament organized a high-profile citizens' conference. The lay panel proposed more stringent regulation and more public funds for agbiotech R&D, as if the latter were benign (Marris 1999). This procedure served to reinforce state-based expertise for managing risks of GM crops and for promoting their innovation. The Environment Ministry took a greater role in risk regulation; the advisory committee was expanded to include more public interest representatives and critics of safety claims (Roy and Joly 2000).

By the late 1990s numerous GM crops were awaiting an EU-wide decision on commercial approval. Some government officials criticized such delays as a threat to 'globalization', while protestors reversed the argument: globalization threatened national sovereignty and democracy. At the June 1999 meeting of the EU Environment Council, many member states declared that they would not consider requests to authorize additional products until new conditions were fulfilled: 'Given the need to restore public and market confidence', among other reasons, the EU must first adopt measures to ensure full traceability and labelling of GM crops across the agro-food chain; risk-assessment procedures must be more transparent and be based on precaution. The EU-wide decision procedure was effectively suspended through a *de facto* moratorium.

The moratorium increased pressures upon the Commission to devise stronger legislation. Eventually the Deliberate Release Directive was revised to include more stringent measures which some member states had already been developing. It provided for time-limited registrations, required market-stage monitoring, and clarified that the risk assessment must consider the effects of any changes in agricultural management methods, such as changes in herbicide usage (EC 2001). Taken together, all these measures incorporated flexible agri-environmental norms, including potential harm to farmland biodiversity from farmer practices. Such reforms potentially enhance public accountability for regulatory judgements – that is, what types of effects should be prevented, what counts as adequate evidence, and thus whether products should be approved. Some national stances were already influenced by wider stakeholder involvement, though such influence was largely limited to regulation.

Diagnoses of legitimacy crisis

Agricultural companies had initially played a central role in setting policy agendas, but protest and commercial blockages against agbiotech opened up the policy process to a wider web of stakeholders. Industry had difficulty in responding to the new context (Levidow *et al.* 2002). The regulatory impasse stimulated policy discussions about 'the public' as a problem.

Many government officials and advisers diagnosed the problem as 'public distrust'. This in turn was attributed to various deficiencies – of public rationality, of public knowledge or risk communication, of government procedures, or all of these (Levidow and Marris 2001). The need to gain or restore trust served as a general rationale to make institutions more trustworthy, through measures which official experts did not always regard as scientifically grounded.

Beyond simply educating the public, proposed remedies included greater public transparency, consultation and even participation, sometimes in the name of 'governance'. Given the credibility problems of 'science-based regulation', 'Science and Governance' was given special prominence as a policy problem, within a broader agenda to overcome the EU's democratic deficit. As these discussions recognized, official expertise was often contested and so could not simply legitimize policy decisions. As a way forward, there were proposals to democratize expertise. According to an official report by that title, official experts and 'counter-experts' often contradict and challenge one another:

> While being increasingly relied upon, however, expertise is also increasingly contested. . . . 'Traditional' science is confronted with the ethical, environmental, health, economic and social implications of its technological applications. Scientific expertise must therefore interact and at times conflict with other types of expertise.
>
> (Liberatore 2001: 6)

At a conference on 'Science and Governance', discussion focused largely on risk assessment rather than R&D policy. Nevertheless, critical perspectives emerged, especially in a workshop on 'Anticipating Risks'. According to the rapporteur:

> The need to involve *normative considerations* in dealing with precautionary-oriented scientific issues is also an element that has a transforming capacity. Many of these issues call for various forms of participatory processes within which stakeholder involvement is important both for the formulation of concepts and questions as well as for the implementation. . . . The broadening of what is really meant by a technology product, including the shift into providing services, *changes the character of innovation characteristics.*
>
> (DG-JRC and Research 2000: 3)

In that vein, agbiotech became a focus of a debate on normative issues (e.g. over how products structure human practices, environmental effects, land use and so on). However, prevalent policy language referred selectively to GM techniques and products as 'the technology', as if more extensive cultivation methods were not a significant innovation.

Conclusions: governing European conflicts over GM crops

As shown in this case study, conflicts over GM crops express divergent views of sustainability, which may be analysed through a tripartite taxonomy (see Table 7.2, by comparison to Table 7.1). In this taxonomy, each view diagnoses problems so as to favour its own concept of what to sustain – for example, different forms of the economy, environment, and society. Each also has different priorities for expertise. Each view may recast key terms, such as Integrated Crop Management, biodiversity, eco-efficiency and community.

The term 'sustainability' has been appropriated by political forces supporting and opposing agbiotech. From a neoliberal view, GM crops offer us eco-efficient solutions to the supposed problem of inefficient agri-inputs, thus potentially intensifying market competition for agri-food foods. From a community view opposing GM crops, more extensive crop protection methods would protect agro-environmental resources as a common good, while 'quality' production would link producers directly with consumers. Such alternatives were counterposed as benign alternatives and as more stringent comparators for evaluating GM crops.

As a form of environmental management, regulatory procedures have mediated between neoliberal and community-type views of sustainability in ways which changed in response to protest. Early on, EU procedures linked environmental protection with a regulatory harmonization which would help to liberalize trade, especially within the EU's internal market. This favoured neoliberal models of the agri-environment as a homogeneous resource for greater productivity. In the late 1990s, protest associated agbiotech with an ominous 'globalization' which would undermine democracy, industrialize agriculture and subordinate farmers to multinational companies.

In response to protest and legitimacy crisis, the EU's technocratic harmonization model gave way to diverse national frameworks for valuing the agri-environment. Risk assessment was extended to protect common goods such as pest-control agents, farmland biodiversity and ground water. As a form of environmental management, regulatory procedures accommodated proposals to evaluate and manage a broader range of plausible undesirable effects from GM crops. Rather than standardize an intensive agricultural model, regulatory procedures could circulate more diverse and stringent criteria across EU member states.

Those changes also involved processes of governing, expressing the need to 'restore public and market confidence' as a collective action problem. Partly with that rationale, official experts acknowledged more scientific uncertainties and potential effects that may warrant regulatory controls. New procedures involved various groups which were sceptical of agbiotech. Some national procedures broadened their expert advisory bodies, sought

Table 7.2 Divergent views of GM crops vis-à-vis sustainable agriculture

View issues	Neoliberal view: high-yield intensification	Environmental management view: precautionary regulation	Community view: extensification
Global problem	genetic deficiencies of crops; inefficient inputs which limit farm productivity	transboundary risks of GM crops; regulatory differences across countries	intensive monoculture; farmer dependence on multinational companies
Concepts of nature	laboratory simulations of biodiversity to protect crops	delicate balance; 'environment' mainly beyond agriculture	biodiversity of cultivars and biocontrol agents
Economic aims	compete for and gain sales of 'green' commodities	avoid trade barriers through common environmental standards	link producers–consumers through quality production
Solution	eco-efficiency replaces energy and materials with genetic information	precautionary measures; biodiversity conservation	less intensive methods of cultivation
Expertise	develop GM crops which reduce agrochemical usage	compare biophysical effects of GM/non-GM crops in advance	develop farmers' knowledge of biodiversity and local resources

means to involve stakeholder groups and established more consultation procedures. For example, the French Parliament organized a citizens' conference, whose lay panel proposed more stringent risk regulation and more public funds for agbiotech R&D. The expert advisory group was expanded to include critical voices. In the UK, environmental groups influenced the design of large-scale experiments testing effects on farmland biodiversity. An additional advisory body was created there with a broad remit to deliberate criteria for environmental harm and sustainable agriculture.

This basis for governing had a relatively greater scope to accommodate dissent and so mediate the conflict, though within limits. Given the EU's treaty obligations, its regulatory procedures could incorporate diverse agri-environmental values only by technicizing them – for example, by devising means to measure biophysical effects of a specific GM product. This task often has methodological difficulties, which generate further disagreements over evidence. Moreover, EU authorities can still limit the definition of harm in practice, so that GM products may still gain approval on a narrow basis.

Another limitation arises from divergent models of agri-societal futures and technological progress. EU innovation policy is still driven largely by an imperative for 'economic competitiveness', generally meaning productive efficiency of intensive monoculture – criteria often in conflict with environmental and product quality. Unless R&D policies are opened up for debate and change, risk regulation will continue to bear the burden of conflicts around divergent sustainability models, without the capacity to promote alternative innovations. Within those limits, processes of governing can only incorporate or marginalize agbiotech critics, who in turn may continue their efforts to undermine public confidence in regulatory procedures.

Concepts for policy and analysis

Finally, this chapter illuminates implicit or contentious meanings of key policy concepts. These are also analytical concepts, though the two usages are often conflated, for example, when social science analysis takes for granted specific policy meanings. This case study highlights ambiguities which warrant analytical attention.

'Globalization' was initially invoked as a dual imperative – of economic competitiveness and treaty obligations – which required approval of GM products. In response, critics identified globalization as a threat – as an imperative to resist undemocratic pressures, to defend sovereignty and to create alternatives. As many NGOs proclaimed, 'Another agriculture is possible'. Thus globalization may be analysed as an ideological construct which reifies policy agendas as external imperatives or threats.

'Sustainable agriculture' too is generally invoked as if it had an obvious meaning. Yet the term is used to promote divergent models of development, while attempting to incorporate or marginalize rival models. These meanings may be analysed to identify contending agri-environmental futures at stake in innovation choices and in regulatory criteria.

'Governance' depends upon a collective action problem which can provide a basis for joint activity by policy actors otherwise in conflict. Governance can displace antagonistic social views on to the arena of risk regulation and public trust, thus providing more subtle ways to legitimize regulatory procedures as trustworthy, or even to legitimize a contentious innovation as progress. Alternatively, it may mean opening up assumptions about the societal problem to be solved by innovation, thus going beyond regulatory conflicts. Thus, whether explicit or implicit, governance may be analysed as an effort to construct and solve a specific problem *as if* it were a collective one.

Acknowledgements

This chapter draws mainly upon a study, 'Safety Regulation of Transgenic Crops: Completing the Internal Market?', funded by the European Commission, DG XII/E5, Ethical, Legal and Socio-Economic Aspects (ELSA), Biotechnology horizontal programme, during 1997 and 1999. Reports available at http://technology.open.ac.uk/cts/srtc/index.html. Research material was provided by our research partners in several member states.

References

Barben, D. (1998) 'The political economy of genetic engineering', *Organization & Environment,* 11, 4: 406–422.

Bové, J. (1998) Speech at the Agen court, 12 February, translated by Greenpeace France, www.greenpeace.org.fr, available at www.cpefarmers.org.

Brundtland, H.G. (1987) *Our Common Future*, Oxford: Oxford University Press.

Camera dei Deputati (1997) *Le biotecnologie*, Atti parlamentari XIII legislatura, Indagini conoscitive e documentazioni legislative n. 5, Roma.

CEC (2001) *A Sustainable Europe for a Better World: A European Union Strategy for Sustainable Development*, COM (2001) 264, available at www.europa.eu.int.

Commission on Global Governance (CGG) (1996) *Our Global Neighborhood*, Oxford: Oxford University Press.

DG-JRC and Research (2000) 'Science and Governance in a Knowledge Society: The Challenge for Europe', 16–17 October, conference conclusions, available at http://www.jrc.es/sci-gov.

Dobson, A. (1996) *Justice and the Environment: Conceptions of Environmental Sustainability and Theories of Distributive Justice*, Oxford: Oxford University Press.

Dreyer, M. and Gill, B. (2000) 'Germany: continued "elite precaution" alongside continued public opposition', *Journal of Risk Research,* 3, 3: 219–226.

EC (2001) European Parliament and Council Directive 2001/18/EC of 12 March on the deliberate release into the environment of genetically modified organisms and repealing Council Directive 90/220/EEC, *O.J.* L 106: 1–38.

EcoSoc (1998) Opinion of the Economic and Social Committee on 'Genetically modified organisms in agriculture – impact on the CAP', *Official Journal of the European Communities*, C 284: 39–50.

EEC (1990) Council Directive 90/220 on the Deliberate Release to the Environment of Genetically Modified Organisms, *Official Journal of the European Communities*, L 117, 8 May: 15–27.

Elert, C. (1991) *Biotechnology at Work in Denmark*, Copenhagen: Danish Board of Technology.

EUREP (1999) Good Agricultural Practice [GAP] Protocol. Euro-Retailer Produce Working Group, available at http://www.eurep.org.

Everard, M. and Ray, D. (1999) *Genetic Modification and Sustainability*. 2020 Vision Series No. 1. Bristol: Environment Agency/Cheltenham: The Natural Step, available at http://www.naturalstep.org.uk/update1a.htm.

FoEE (1996–1998) *FoEE Biotech Mailout*, Brussels: Friends of the Earth Europe Biotechnology Programme, available at www.foeeurope.org/biotechnology.

FoEE (2000) 'EU environment ministers discuss GMO approvals', *FoEE Biotech Mailout* 6, 5: 4, available at www.foeeurope.org/biotechnology/about.htm.

Ford, L. (2003) 'Challenging global environmental governance: social movement agency and global civil society', *Global Environmental Politics,* 3, 2: 120–134.

Gill, B. (1993) 'Technology assessment in Germany's biotechnology debate', *Science as Culture,* 4, 1: 69–84.

Gill, B. (1996) 'Germany: splicing genes, splitting society', *Science and Public Policy,* 23, 3: 175–179.

Greens/EFA (2001) *Agroecology: Toward a New Agriculture for Europe*, Brussels: Greens/EFA in the European Parliament.

Haerlin, B. (1990) 'Genetic engineering in Europe', in P. Wheale and R. McNally (eds) *The BioRevolution: Cornucopia or Pandora's Box?*, London: Pluto.

Heller, C. (2002) 'From scientific risk to *paysan savoir-faire*: peasant expertise in the French and global debate over GM crops', *Science as Culture,* 11, 1: 5–37.

Imhof, H. (1998) 'Challenges for sustainability in the crop protection and seeds industry', talk by Novartis officer at Rabobank International's Global Conference.

Krimsky, S. and Wrubel, R. (1996) *Agricultural Biotechnology and the Environment: Science, Policy and Social Issues*, Chicago: University of Illinois Press.

Levidow, L. and Bijman, J. (2002) 'Farm inputs under pressure from the European food industry', *Food Policy,* 27, 1: 31–45.

Levidow, L. and Carr, S. (eds) (2000) 'Precautionary Regulation: GM Crops in the European Union', special issue of the *Journal of Risk Research*, 3, 3: 187–285.

Levidow, L. and Marris, C. (2001) 'Science and governance in Europe: lessons from the case of agbiotech', *Science and Public Policy,* 28, 5: 345–360.

Levidow, L., Carr, S., von Schomberg, R. and Wield, D. (1996) 'Regulating agricultural biotechnology in Europe: harmonization difficulties, opportunities, dilemmas', *Science and Public Policy,* 23, 3: 135–157.

Levidow, L., Oreszczyn, S., Assouline, G. and Joly, P-B. (2002) 'Industry responses to the European controversy over agricultural biotechnology', *Science and Public Policy*, 29, 4: 267–275.

Liberatore, A., rapporteur (2001) 'Democratising Expertise and Establishing Scientific Reference Systems', Report of the Working Group 1b, Broadening and enriching the public debate on European matters, *White Paper on Governance*, available at http://europa.eu.int/comm/governance/areas/group2/index_en.htm.

McKechnie, S. (1999) 'Food fright', *Guardian*, 10 February (Director of the Consumers Association).

Magretta, J. (1997) 'Growth through global sustainability: an interview with

Monsanto's CEO, Robert Shapiro', *Harvard Business Review,* January to February: 79–88.

Marris, C. (1999) 'Between consensus and citizens: public participation in technology assessment in France', *Science Studies,* 12, 2: 3–32.

Monsanto (1997) *Report on Sustainable Development,* St Louis, MO: Monsanto Company.

Newell, P. (2003) 'Globalization and the governance of biotechnology', *Global Environmental Politics,* 3, 2: 56–71.

Novartis (1998) *Novartis Bt-11 Maize,* Basel: Novartis, available at www.novartis.com.

Paterson, M., Humphreys, D. and Pettiford, L. (2003) 'Conceptualising global environmental governance: from interstate regimes to counter-hegemonic struggles', *Global Environmental Politics,* 3, 2: 1–10.

Roy, A. and Joly, P-B. (2000) 'France: broadening precautionary expertise?', *Journal of Risk Research,* 3, 3: 247–254.

Sachs, W. (ed.) (2003) *The Jo-Burg Memo: Fairness in a Fragile World. Memorandum for the World Summit on Sustainable Development,* Berlin: Heinrich Böll Foundation, available at www. joburgmemo.org.

Schmidheiny, S. (1992) *Changing Course: A Global Business Perspective on Development and the Environment,* Cambridge, MA: MIT Press.

Terragni, F. and Recchia, E. (1999) 'Italy: Precaution for Environmental Diversity?', Report for 'Safety Regulation of Transgenic Crops: Completing the Internal Market', DGXII RTD project coordinated by the Open University, available at http://technology.open.ac.uk//cts/srtc/index.html.

Todt, O. and Lujan, J. (2000) 'Spain: commercialisation drives public debate and precaution', *Journal of Risk Research,* 3, 3: 237–245.

Toft, J. (2000) 'Denmark: potential polarization or consensus?', *Journal of Risk Research,* 3, 3: 227–236.

Torgerson, H. and Seifert, F. (2000) 'Austria: precautionary blockage of agricultural biotechnology', *Journal of Risk Research,* 3, 3: 209–217.

Woodhouse, P. (2000) 'Environmental degradation and sustainability', in T. Allen and A. Thomas (eds) *Poverty and Development into the 21st Century,* Oxford: Oxford University Press, in association with the Open University, Milton Keynes.

Young, O.R. (1994) *International Governance: Protecting the Environment in a Stateless Society,* Ithaca, NY: Cornell University Press.

8 Governing agriculture through the managerial capacities of farmers

The role of calculation

Vaughan Higgins

Introduction

In recent years, agricultural industries in both developed and developing nations have been transformed. The broad literature on global agri-food restructuring represents an attempt by scholars to come to terms with the regulatory nature, scale and consequences of these changes (see Bonanno *et al.* 1994; Burch *et al.* 1999; Burch *et al.* 1996; Goodman and Watts 1997; Le Heron 1993; McMichael 1994; Marsden *et al.* 1990). Characteristic of much work on agricultural 'restructuring' is the argument that farmers are placed under increasing pressure by global social forces and actors, and must improve their productivity and efficiency or leave the industry. Drawing on a critical political economy approach, this literature is influential in focusing attention on the macro-level global forces that shape on-farm practices, resulting in a loss of agency by smaller scale commodity producers to the profit-making interests of multinational agribusinesses. Accordingly, farmers are seen to have little choice but to conform to these global neo-liberalist regulatory arrangements.

The critical political economy approach to regulation has had considerable influence in the agri-food restructuring literature (Higgins 2001; see also Buttel 2001). However, while a political economy approach provides a broad macro-structural explanation of the causes of change, and the consequences for the regulation of farming practice, it is unable to account adequately for *how* this occurs (see Higgins 2002a). In its search for the hidden dynamics and actors driving change, it could be argued that a political economy perspective relies on a sovereign model of power, thus overlooking the more subtle and productive 'arts of government' (Foucault 1991) that are increasingly evident in globalization processes (Larner and Walters 2002). Neglected in particular are the specific and localized practices of governing through which farmers are encouraged to become more efficient and productive (see Le Heron 2003).

The purpose of this chapter is to broaden the agenda of a political economy analysis by demonstrating the significance of productive forms of agricultural regulation in shaping the capacities of farmers as modern businesspeople engaging in 'efficient' and 'competitive' practices. Drawing on a Foucauldian-inspired 'analytics of governmentality', and supported with conceptual tools from the sociology of science and technology, I argue that calculation has assumed major prominence in the constitution of farmers' agency, and particularly in the governing of their managerial capacities. While calculation has long been important in farm management, only recently has it emerged as a key vehicle of advanced liberal agricultural governing. Rather than power being exercised over farmers, a governmentality approach serves to direct attention to the practices that enable farmers' managerial capacities to be rendered visible as objects of power-knowledge, and thereby made amenable to intervention. Moreover, I argue that the governing of these capacities requires the building of centres of calculation through which the conduct of farmers may be shaped at a distance. I draw upon a case study of a dairy planning workshop operating in Victoria, Australia to illustrate the significance of calculation as a key technology of contemporary agricultural governing.

A governmentality perspective on the governing of agriculture

An analytics of governmentality provides a strong conceptual basis for exploring the productive nature of modern agricultural governing. The governmentality literature is based on the assumption that governing cannot be reduced to a singular actor or logic such as the state or the profit-making logic of capital. Such a focus is of particular significance in examining projects of agricultural regulation in that it enables closer attention to the 'surfaces, practices and routines' (Larner and Walters 2002: 2) that assemble globalization, rather than the frequent tendency in the agricultural restructuring literature to 'fetishize the global' (Buttel 1996: 32; see also Buttel 2001). Based upon a Foucauldian 'micro-physics' of power, regulation is instead conceptualized as an effect of heterogeneous and shifting discursive and material relations, in which the capacities and limits of governing, and the governed, are constituted. In this sense, the problem of power is reformulated as 'not so much a matter of imposing constraints on citizens as of "making up" citizens capable of bearing a kind of regulated freedom' (Rose and Miller 1992: 174). A concise starting point for such an approach involves the identification and analysis of specific situations where governing is called into question or problematized.

Problematizations of rule are at the heart of a governmentality analysis. Governing is understood as a problematizing activity that is 'intrinsically linked to the problems around which it circulates, the failings it seeks to rectify, the ills it seeks to cure' (Rose and Miller 1992: 181). Programmes of governing are characterized by objectives and strategies seeking to provide answers to various problems. Rather than focusing on the implementation of these 'solutions', as measured against some ideal of rule, the task of the researcher is to reconstruct the problematizations that form the basis of programmes and strategies. As Rose notes, the reconstruction of problematizations is an important means of according programmes intelligibility as answers, and in 'enabling their limits and presuppositions to be opened for interrogation in new ways' (Rose 1999: 58). This is of particular significance to analyses of agricultural governance in that it shifts the focus from explaining the hidden dynamics behind ostensibly pre-constituted projects of rule, to questions of how these projects are rendered knowable and governable in the first place (Higgins 2002a).

Problematizations of governing are analysed in terms of what are called *rationalities* and *technologies*. Rationalities of governance comprise a discursive means by which the problems that form the basis of rule are reflected upon and linked together in coherent ways. They are a crucial part of making governing thinkable and practicable, and in defining the proper functions, limits and capacities for rule to be operable (Rose 1999). Governing, from a govenmentality perspective, is based also on what Dean (1999: 31) calls the 'technical means' that prescribe how governing may be achieved. This involves an examination of not only rationalities of rule, but also the political technologies through which governing is deployed and seeks to achieve effects. In order for rationalities of rule to be transformed into a means for shaping conduct, technical devices need to be deployed 'that render a realm into discourse as a knowable, calculable and administrable object' (Miller and Rose 1990: 5). According to Rose and Miller (1992: 183), these technologies are heterogeneous and include:

> the humble and mundane mechanisms by which authorities seek to instantiate government: techniques of notation, computation and calculation; procedures of examination and assessment; the invention of devices such as surveys and presentational forms such as tables; the standardization of systems for training and the inculcation of habits; the inauguration of professional specialisms and vocabularies; building designs and architectural forms.

It is here that calculation assumes particular significance. In order to shape human conduct according to political objectives, authorities must ensure that strategies formulated at one location (e.g. state agencies, farm planning workshops) can be 'translated'[1] (Rose 1999) into action at another. This is by no means a straightforward process and is contingent on the building of what Rose and Miller (1992: 185), drawing on the work of Latour, call 'centres of calculation' through which reality can be made stable, mobile, comparable, combinable, and represented in a form in which it can be debated and diagnosed. In the work of Latour, inscription devices such as maps, charts, drawings and diagrams represent an important means for events, places and people to be governed at a distance. To act at a distance these inscriptions must be rendered *mobile*, so that they may be brought back; *stable*, so that they may be moved back and forth without additional distortion, corruption or decay; and *combinable*, so that they may be accumulated or aggregated (Latour 1987: 223). Inscription, therefore, makes the building of centres of calculation possible through the representation of practices in a mobile, stable and combinable form. If these centres remain durable, they ensure that certain calculations and practices are valorized over others 'by means of a passage through the centre' (Rose and Miller 1992: 189).[2]

Building farmers' capacities: governing in an advanced liberal way

In what ways do the above analytical tools apply to contemporary ways of governing the conduct of farmers? Through the application of a governmentality perspective, a number of scholars have noted in recent policy initiatives the emergence of a rationality of rural governing aimed towards facilitating the development of active citizens who take greater responsibility for their self-governance (e.g. Herbert-Cheshire 2000; Higgins 2002b; Ward and McNicholas 1998). This involves a shift away from state-based 'welfarist' forms of intervention to a focus upon governing economic life through the calculative capacities of individuals – referred to by governmentality scholars as governing in an advanced liberal way (Dean 1999; Rose 1993, 1999). For advanced liberal forms of rule, 'economic government is to be de-socialized in the name of maximizing the entrepreneurial comportment of the individual' (Rose 1999: 144).

Contemporary programmes for the governing of farming practices are consistent with attempts to govern in an advanced liberal way – their explicit aim is to provide a platform through which the planning and managerial capacities of farmers may be enhanced to improve agricultural efficiency and sustainability. Rather than providing government assistance to farmers, programmes increasingly have an explicit focus on bringing about long-term attitudinal and cultural change in farmers' conduct (e.g. Higgins 2002b; Kidd *et al.* 2000; Pretty and Chambers 1994; Winter 1997). Thus such rationalities accord with attempts to create what Dean (1999) calls 'active citizens' who are equipped with the capacity to conduct themselves in a 'responsible' entrepreneurial way. Equally, those who do not make so-called responsible choices are seen as 'targeted populations' (Dean 1999), high-risk groups who lack the necessary life management skills (Higgins 2002b). This focus on enterprising conduct requires the strengthening of economic citizenship through marketized technologies that seek to enhance the capacities of individuals – what Dean (1999) refers to as 'technologies of agency'.

Programmes seeking to build the calculative capacities of farmers have only recently been examined from a governmentality perspective. The work of both Martin (1997) and Lockie (1999) provides a relevant starting point in reflecting on how programmes aimed at improving farm productivity and sustainability seek to govern in an advanced liberal way. These writers focus on Australia's National Landcare Programme, a participatory strategy of natural resource management involving a partnership of farmers and government. What the work of both Martin and Lockie illustrate in the context of this chapter is not only the significance of farmers' capacities as sites of governing, but also the key role of calculation in this process. Programmes such as Landcare that seek to 'empower' farmers to become better (in this case environmental) managers are rendered operable through advanced liberal rationalities and technologies of rule that encourage the building of self-calculating capacities. Educational technologies are, particularly for Martin (1997), crucial in constituting farming practices in an advanced liberal way, and in attempting to act at a distance to align the objectives of authorities with the conduct of farmers. However, neither Martin nor Lockie explore the characteristics of such technologies, how they are deployed in specific contexts, or the ways in which they govern through farmers' capacities and constitute them as calculative agents. These issues are addressed in the following section of the paper where I explore a farm planning workshop in the Australian dairy industry and the technologies of calculation that seek to constitute and shape the practices of participating farmers.

Advanced liberalism and the calculation of dairy farm planning

The Australian dairy industry represents an ideal site to explore how the managerial capacities of farmers are governed in an advanced liberal way. Dairying in Australia is divided into two main markets: a 'white' fresh milk market and a 'yellow' milk market for manufactured dairy products. Over 63 per cent of total milk yield is produced in Victoria, with this state also using most of its milk for manufacturing purposes (Cocklin and Dibden 2002: 30). Since domestic consumption of dairy products grows only slowly, the development of the industry is, both in recent times and in the future, dependent on the supply of low-cost products such as milk powder, cheese and butter to the expanding Asian market (ABARE 2003). In fact, Australia now exports over half of its total dairy production, making it the third largest exporter of dairy products behind New Zealand and the European Union.

Declining terms of trade, along with the increasingly export-oriented nature of the industry, has meant an ongoing need for improvements in productivity and efficiency by producers. Those producers unable or unwilling to adjust the structure of their operations have tended to leave the industry. For instance, between 1974 to 1975 and 2001 to 2002 the number of dairy farms in Australia declined from approximately 30,000 to under 11,000 (ABARE 2003). While farm numbers have declined, the industry has experienced substantial increases in productivity through larger herd sizes, higher milk yield per cow and the expansion of existing farms. This indicates that those farmers remaining have been forced to find ways to increase output to keep up with rising input costs.

Up until early 2000, farm-gate milk prices were regulated at the Australian state level 'with different subsidies paid according to the end use of the milk' (Cocklin and Dibden 2002: 31). This meant that milk prices remained relatively stable. However, this regulatory system was subjected to increasing levels of scrutiny from the late 1990s. The Federal government, along with many farmers' organizations, argued that continued regulation was contrary to the National Competition Policy,[3] and that it weakened Australia's bargaining position for trade liberalization in World Trade Organization negotiations (Cocklin and Dibden 2002: 31). Such arguments in favour of dairy reform[4] led to farm-gate milk prices being deregulated completely in July 2000.

Since deregulation of the industry, there has been pressure on farmers to adopt improved planning techniques as a means to enhance their productivity in a highly competitive market environment. Deregulation highlighted for many in the industry that dairy farming was a business requiring sound planning techniques in order to respond to commercial pressures (Cocklin and Dibden 2002). While numerous farm management courses existed prior to this shift, deregulation undoubtedly placed additional scrutiny on farmers' managerial practices, in particular their capacity to adjust not simply to a longer term cost-price squeeze, but now fluctuating farm-gate milk prices. If farmers wanted to earn a reasonable income, and to avoid exit from the industry, it was imperative to change the way in which they managed their enterprise. Training in farm management and planning represented a crucial means for this to be achieved.

Training in dairy farm planning and management has occupied a prominent place on the agenda of state departments of agriculture for a number of years. Prior to deregulation, a developing emphasis was evident in dairy programmes on activities seeking to build human resources rather than 'top-down' technology transfer. In part, this reflects changing views about government's role in the economy and particularly an (advanced liberal) desire for reduction in government spending (Marsh and Pannell 2000: 606). However, there is also a broader philosophical shift towards initiatives that seek to *facilitate* change through education and training activities. In many ways, the desire for reduced public spending and a changing extension philosophy are complementary. If programmes equip farmers successfully with the skills to plan for and manage their enterprises over the longer term, ongoing government support will not be needed; these farmers will effectively be reliant on their own capacities (see Higgins 2002b).

National initiatives such as the National Landcare Programme (NLP) and the Property Management Planning Programme (PMP) paved the way for a 'bottom-up' approach to farm planning and management. This focus on building 'active' farmers is evident in dairy extension projects such as the highly successful Victorian-based programme, Target 10® (see Boomsma 2000), which has shaped how other dairy extension and training initiatives are run. Programmes such as Target 10®, which have a strong focus on partnership and participation as a means to improve management practices, tend to assume that 'farmers' inaction is caused more by a lack of shared understanding of the problem than lack of awareness of scientists' solutions' (Marsh and Pannell 2000: 611). Hence, the core objective is to create a learning environment in which farmers can reflect upon their practices in a way that provides the basis for change in those practices.

The following example outlines a programme established to enhance the managerial and planning capacities of dairy producers. Observations of the programme form part of a larger project conducted by the author into the role of computer-based planning technologies in the governance of dairy farming practices. Data on how the programme, and the software that forms the centre-piece of it, have shaped the practices of dairy producers will be the subject of future studies. The focus here will be primarily on the rationalities and technologies that provide the basis for problematizing the conduct of farmers and seeking to govern them as calculable agents.

The dairy planning programme

This section of the chapter examines a dairy planning programme (DP)[5] operating in the state of Victoria, Australia. The DP programme is a means of assisting dairy farmers in farm resource planning and allocation, and comprises a two-day training workshop that aims to allow farmers to 'test their options and rapidly set up a comprehensive annual dairy profit plan that they have confidence in' (Course Training Notes). The programme is privately operated, consistent with the general trend in agricultural extension and training (Kidd *et al.* 2000; Marsh and Pannell 2000), but between 50 and 75 per cent of the cost of participation is subsidized by FarmBis, the core Federal programme for promoting improved on-farm management practices. Since its commencement in 2001 to the time of writing (May 2004), the programme has attracted almost 600 participants in over seventy-five workshops, which have been conducted not simply in Victoria but also in many parts of eastern Australia. For a relatively small and privately operated training course this is an impressive figure, and shows considerable demand among dairy farmers, in a recently deregulated industry, for initiatives seeking to build their planning capacities.

The collection of data discussed in the remainder of the chapter was obtained through observations by the author at two workshops, and two rounds of in-depth interviews with ten farm management teams[6] who had participated in the workshop in the eastern Victorian region of Gippsland. The operations of these management teams varied considerably, ranging from 190 to 500 cow herds, and a farm area of between seventy-five and 150 hectares devoted to dairying.

As noted above, the workshops aim to provide farmers with the capabilities to set up an annual dairy profit plan. From a governmentality perspective, the workshops are based on a certain problematization of farming practices, and how these may be governed most effectively. In addition, the workshops seek to construct and mobilize an advanced liberal representation of the entire dairy industry that targets the conduct of individual farmers as the legitimate site of change. This problematization may be characterized as follows. In a deregulated environment many farmers are facing pressure to remain viable; much of this pressure is a result of a fixed mindset and lack of knowledge on the part of farmers concerning how to run their farms as a business; since numerous farmers have already left the industry, those remaining need to find ways to avoid the same fate; on this basis the workshop provides a solution to these problems by training farmers in planning techniques that will enable them to not only remain in the industry, but also prosper and live a comfortable lifestyle underpinned by a 'reasonable' milk price (Higgins and Kitto 2004). The techniques of calculation promoted via the workshops are therefore constructed as central in farmers assisting farmers in acquiring the capacities to make a business profit. Viewed in this context, the workshops need to be viewed as more than simply a localized form of regulation. The manner in which they problematize Australian dairy farming renders the programme part of a much broader advanced liberal governmental rationality where farmers are argued to require playing an active role in their self-governing (Higgins 2001).[7] Similar to other farm planning and management courses in Australia, this type of problematization of farmer conduct is relatively common (see e.g. Higgins 2002b). However, what is less common is the use of a specific computer software package that forms the centre-piece of the workshop. There are three main technologies associated with the use of the computer software through which farmers are encouraged to think of themselves and their practices in a calculable way: (1) an electronic questionnaire; (2) a profit calculator; and (3) printouts of farmers' calculations.

The electronic questionnaire

Farmers who enrol in the two-day training are expected to complete an electronic questionnaire a number of days prior to the workshop requiring them to enter data on such details as their farm finances, milk production, calving patterns, physical characteristics of the farm, and resource and input quantities. These 'raw' data form the foundation of subsequent farm planning and resource allocation. In this way, the questionnaire and, more broadly the computer software, is a technology of government that attempts to construct a 'centre' (Latour 1987; Miller and Rose 1990) which farmers must use as the basis of their calculations in order to complete the training successfully and, in the longer term, improve their planning practices. The questionnaire not only enables particular farming practices to be accumulated in statistical terms and brought 'inside' the software, but also at the same time constitutes those practices as representative of the 'outside' farming enterprise. This creates an equivalence (see Berg 1997; Callon 1986) in which the computer representation effectively becomes 'the farm' that is used subsequently as the basis of informing and shaping future action.

Simply by completing the electronic questionnaire, farmers are enrolled into governmental practices that seek to break down and categorize their farm inputs and outputs into standard quantifiable dimensions which may be used as the basis for judging and manipulating planning outcomes. While the broad categories used as part of the questionnaire appear familiar to most farmers undertaking the course, the level of detail required means that in many instances farmers are forced to estimate figures so that their farm data are aligned with the categories in the software. As one farmer noted, 'they tried to split the costs of rearing replacements separately, whereas I haven't bothered to do that within our financial system. So some of those I had to make estimates' (F#6). Another observed:

> it's . . . [repairs and maintenance] . . . just one problem, but they want to know repairs for the dairy, repairs for machinery, repairs for the out paddock, blah, blah, blah. And I mean it was really just a guess of some of those figures as long as it reconciled with your total columns, so you didn't distort the figures (F#8).

This might be seen as a process of 'configuring the user' (Woolgar 1991) through the deployment of standards. According to Star (1991), technological standardization is a crucial part of the stabilization of socio-technical networks. Standardization of farm data according to specific categories is therefore a crucial first step in achieving universality of practices (see Latour 1987) and, in this case, getting farmers to reflect on their practices in a similar way. Nevertheless, as Barry (2001: 63) points out, far from being fixed and unchangeable, standardization is an ongoing and inherently problematic process that creates 'new sites and objects of political conflict'. The estimation by farmers of some farm figures that occurs as a consequence of standardization seems to illustrate the fragility of standards and the possibility for alternative calculative practices to exist alongside those prescribed in the workshops.

The profit calculator

The standardization of farm data to accord with the categories in the software represents one way in which calculation forms the basis of how farmers reflect on their conduct. However, what is of most technological significance in the software is a function that enables the figures entered into each category to be combined and manipulated by farmers. The profit calculator function allows farmers to view farm resource allocation and profit for the previous year and use this as the basis for formulating 'what if?' scenarios for the following year. Farmers can manipulate data on such factors as resources, production, herd energy needs and feed inputs to test whether a given production strategy is likely to be profitable. From the perspective of farmers, this allows production strategies to be tested in the 'virtual' world before being applied in the 'real' world. The manipulation of statistics in this way may be seen as simply a means of improving control over the farm enterprise. However, from a governmentality perspective, it renders 'the farm' and farming practices visible, and amenable to intervention, in new types of ways (Murdoch and Ward 1997).

The combinability of the statistics is a means of enrolling farmers into the network-building activities of the workshop by making planning possibilities visible that may not have previously been considered. Visualization of present production patterns through the software gives an immediate indication of the farm's position and enables future planning in line with what the statistics 'say'. Farmers who were interviewed as part of research for this chapter found the combinability of production scenarios to be a particularly valuable feature of the software in enabling them to see 'what the results will be, rather than sort of waiting for it to happen, or otherwise just you judging that that may be the outcome' (F#3) and having 'far more of an idea where you're heading or where you're trying to head' (F#11). Many participants felt that a key advantage of the software was its accuracy to the extent that 'you'll be able to see straight away things you're doing and things you're not doing. And generally the path you've been taking sometimes, you might have thought was the right way, but as soon as you have it sitting out black and white you can see straight away that it's not' (F#13). Thus, the statistical calculations made possible by the software configured the user (Woolgar 1991) in an explicit way by opening up farm practices to scrutiny and providing the basis for farmers to change their conduct to accord with what the statistics indicated was the most profitable outcome.

Printouts from the workshop

Over the two days of the workshop, two to three trainers introduce approximately ten to fifteen management teams[8] to the planning techniques contained in the software. The trainers assist producers in becoming acquainted with the software and manipulating individual farm resource data in order to produce profitable farm outcomes. Throughout the workshop, farmers are permitted to print any calculations they make as the basis of future resource planning. It is expected that farmers take the printouts produced in the course of the workshop back to their farms and use these as the basis of profit planning in the following year. The printouts detail the farmers' dairy farm profit plan devised in the course of the workshop and include statistics on: pasture use assessment; customized profit scenarios at different milk and feed prices; customized annual feeding and financial budgets; customized capital works and debt-servicing budgets; a cash feasibility report; a risk check assessment of the profit plan; a customized daily milk map linked to the profit plan; and a customized herd map linked to the calving pattern (Dairy Planning Promotional Notes). Significantly, the printouts enable what Miller and Rose (1990), drawing on Latour, term 'action at a distance' in which the planning principles inscribed in the software are rendered mobile and deployed to shape the decision-making capacities of farmers beyond the context of the workshop. In other words, even though the software package may not be used physically by farmers beyond the confines of the workshop, the calculations it makes possible may be reproduced in paper form and thereby act as a highly mobile centre of calculation.

The printouts are mobilized in a way that (re)presents various dimensions of farming practices as problem sites that require individual action in order to produce a profitable enterprise. Thus farmers are encouraged to reflect on planning practices in a way that is consistent with representations detailed on the printouts. As a consequence, the printouts assume prominence as the primary material inscription within a centre of calculation (the workshop itself) that benchmarks the practices and techniques requiring attention to run a profitable dairy farm. Of particular importance is the 'responsibilization' of the farmer through this governing at a distance process which reduces the success or failure of the plan to individual planning competencies. It is assumed that farmers should be in control of their plans and, if problems are experienced, to scrutinize the quality and accuracy of the original inputted figures. In fact, interviews with farmers suggest that even slightly 'imprecise' farm data have the capacity to produce a plan that the software 'indicates' is not going to be workable in practice. For instance:

> You've got to make sure you've got everything, all the information that's required to go into the programme, sorted and in its appropriate place and exact before you go into it. The first course that we did, we weren't that prepared, and ah, we were sitting there the first day working things out and, fair enough, we would have got to within 5 or 6 per cent, like we would have bugger all out, but we were still out. We weren't exact. So, you know, it could have been that, 5 or 6 per cent out now, 5 or 6 per cent out in something else, at the end of the day it could have added to a big bloody blow out (F#13).

> I had a few problems there because I double dipped on a few costs, and I actually couldn't make it work. And ah [one of the trainers] came around, and he'd go 'oh gee you know, you'll have to give dairying away.' And I thought 'gee,' I had to get something like about over 7,000 litres per cow to make it break even because the costs were too high. And I know what happened is that I had my costs and my share farmers' costs so it was doubled up, so a few of them were double dipping but only because I'd separated everything and just did it on my

business, and they said 'no, no, no, no you have to do the whole business and put the share farmers share in as imputed labour,' but then I went 100 per cent cost for us to put all their share plus their costs in labour so it just went through the roof. So that was, I was really thinking, I was thinking 'oh how can I do this?' (F#7)

Farmers are thus encouraged to calculate and to adjust, in an iterative fashion if necessary, their inputs for the following year until they produce a plan that the software indicates is workable. In the case of the second farmer quoted above, in particular, this places pressure on producers to divorce the statistical calculations they make from their previous learned experience or 'local' knowledge in the planning process.

However, the printouts are by no means immutable. While they may facilitate the shaping of conduct at a distance, inscriptions can also be ignored, 'incorrectly' deployed, or problematized by farmers in unanticipated ways (see Higgins and Kitto 2004). For instance, the production plan detailed on the printouts may be compromised if the cost of inputs changes, or if there are unanticipated expenses due to drought conditions. Equally, farmers may lose the printouts or profess to be too busy engaged in other work to refer to them. Technical devices can enact political programmes at a distance, but they can also act in contradiction to the intended political objectives for which they are deployed (Barry 2001; Callon 1991; Michael 2000). The next stage of the research project seeks to examine the contingency, messiness and 'subterranean translations' (O'Malley 1996) through which farmers negotiate the calculative techniques inscribed in the printouts and software into their practices.

Conclusions

This chapter demonstrates the significance of productive forms of governing in contemporary programmes of agri-food regulation, and the role of calculation in this process. An analytics of governmentality enables agri-food researchers to explore the less obvious rationalities and technologies that attempt to constitute and shape the capacities of farmers, rather than commencing an analysis of regulation in terms of macro-sociological 'global' drivers of change. In moving towards examining the *how* of agricultural governing, I have argued that more attention needs to be paid to technologies of calculation that render certain farming practices visible, and, through statistical representations, encourage farmers to reflect on their management and planning in an advanced liberal way. The dairy planning course outlined in this chapter provides an insight into how mechanisms of agricultural governing seek to target the managerial conduct of farmers, and to build their capacities as calculable agents. The computer software deployed as part of this course problematized farmers' existing capabilities and encouraged them to act on their managerial conduct in new ways. Statistics were central in problematizing existing practices and in 'showing' what forms of conduct needed changing in order to achieve a business profit. The printouts from the course acted as a highly mobile centre of calculation allowing the plans formulated through the software to be taken away and applied by farmers.

While only a preliminary exploration of the role of calculation as a key technology of modern agricultural governing, this chapter signals two useful lines of inquiry requiring future attention by agri-food scholars. First, the shift towards 'bottom-up' forms of agricultural regulation requires additional analytical tools to those provided by political economy. Farmers' capacities can no longer be examined as just a response to global forms of regulation, or as a form of resistance. As this chapter demonstrates, the capacities of farmers form a key *vehicle* through which agriculture is governed. In constituting farmers as calculable agents, advanced liberal technologies of rule seek to link the concerns of farmers with broader governmental objectives. In this way, a focus on localized forms of governing does not necessarily mean the neglect of 'global' regulation. Second, I draw attention to the role of computer-based planning and management systems as a productive form of governing. New technologies are clearly important both in constituting and shaping the managerial capacities of farmers. As more and more farmers adopt computers for management and planning purposes, further research will be required on how these software packages both govern, and are applied and translated by, farmers.

Acknowledgement

The research on which this chapter is based is funded by a Monash University Small Grant.

Notes

1 The notion of translation is adapted by Rose from the work of Michel Callon and Bruno Latour. Rose (1999: 48) defines translation as the alignments 'forged between the objectives of authorities wishing to govern, and the personal projects of those organizations, groups and individuals who are the subjects of government'.
2 See Murdoch and Ward (1997) for the constitutive role of statistics as a centre of calculation in the emergence of a British farming 'sector'.
3 All states had signed on to the National Competition Policy by the late 1990s.
4 Discussed in more detail by Cocklin and Dibden (2002).
5 For reasons of confidentiality the name of the dairy planning course, and associated workshop tools, has been changed.
6 Four of these management teams were couples. The other interviews were conducted with only one member of the management team.
7 See Dean (1999) for a more general discussion of this point.
8 These consist predominantly of husband-and-wife teams. However, a small number of male farmers only attend the training.

References

Australian Bureau of Agricultural and Resource Economics (ABARE) (2003) *Australian Dairy Industry: Productivity and Profit*, Canberra: Dairy Australia.

Barry, A. (2001) *Political Machines: Governing a Technological Society*, London: The Athlone Press.

Berg, M. (1997) *Rationalizing Medical Work: Decision-support Techniques and Medical Practices*, Cambridge, MA: MIT Press.

Bonanno, A., Busch, L., Friedland, W.H., Gouveia, L. and Mingione, E. (eds) (1994) *From Columbus to ConAgra: The Globalization of Agriculture and Food*, Lawrence: University of Kansas Press.

Boomsma, J. (2000) *Target 10: Empowering Dairy Communities to Manage Change*, Melbourne: Department of Natural Resources and Environment.

Burch, D., Goss, J. and Lawrence, G. (eds) (1999) *Restructuring Global and Regional Agricultures: Transformations in Australasian Agri-food Economies and Spaces*, Aldershot: Ashgate.

Burch, D., Rickson, R. and Lawrence, G. (eds) (1996) *Globalization and Agri-food Restructuring: Perspectives from the Australasia Region*, Aldershot: Ashgate.

Buttel, F. (1996) 'Theoretical issues in global agri-food restructuring', in D. Burch, R. Rickson and G. Lawrence (eds) *Globalization and Agri-food Restructuring: Perspectives from the Australasia Region*, Aldershot: Ashgate.

Buttel, F. (2001) 'Some reflections on late twentieth century agrarian political economy', *Sociologia Ruralis*, 41, 2: 165–181.

Callon, M. (1986) 'Some elements of a sociology of translation: domestication of the scallops and the fishermen of St. Brieuc Bay', in J. Law (ed.) *Power, Action and Belief: A New Sociology of Knowledge?*, London; Routledge & Kegan Paul.

Callon, M. (1991) 'Techno-economic networks and irreversibility', in J. Law (ed.) *A Sociology of Monsters: Essays on Power, Technology and Domination*, London; Routledge & Kegan Paul.

Cocklin, C. and Dibden, J. (2002) 'Taking stock: farmers' reflections on the deregulation of Australian dairying', *Australian Geographer*, 33, 1: 29–42.

Dean, M. (1999) *Governmentality: Power and Rule in Modern Society*, London: Sage.

Foucault, M. (1991) 'Governmentality', in G. Burchell, C. Gordon and P. Miller (eds) *The Foucault Effect: Studies in Governmentality*, London: Harvester Wheatsheaf.

Goodman, D. and Watts, M. (eds) (1997) *Globalising Food: Agrarian Questions and Global Restructuring*, London: Routledge.

Herbert-Cheshire, L. (2000) 'Contemporary strategies for rural community development in Australia: a governmentality perspective', *Journal of Rural Studies*, 16, 2: 203–215.

Higgins, V. (2001) 'Calculating climate: "advanced liberalism" and the governing of risk in Australian drought policy', *Journal of Sociology*, 37, 3: 299–316.

Higgins, V. (2002a) *Constructing Reform: Economic Expertise and the Governing of Agricultural Change in Australia*, New York: Nova Science.

Higgins, V. (2002b) 'Self-reliant citizens and targeted populations: the case of Australian agriculture in the 1990s', *ARENA Journal*, 19: 161–177.

Higgins, V. and Kitto, S. (2004) 'Mapping the dynamics of new forms of technological governance in agriculture: methodological considerations', *Environment and Planning A*, 36: 1397–1410.

Kidd, A.D., Lamers, J.P.A., Ficarelli, P.P. and Hoffman, V. (2000) 'Privatising agricultural extension: caveat emptor', *Journal of Rural Studies*, 16, 1: 95–102.

Larner, W. and Walters, W. (2002) 'Globalization as governmentality', Paper presented at the ISA XV *World Congress of Sociology*, Brisbane, 7–13 July.

Latour, B. (1987) *Science in Action: How to Follow Scientists and Engineers through Society*, Cambridge, MA: Harvard University Press.

Latour, B. (1990) 'Drawing things together', in M. Lynch and S. Woolgar (eds) *Representation in Scientific Practice*, Cambridge, MA: MIT Press.

Le Heron, R. (1993) *Globalized Agriculture: Political Choice*, Oxford: Pergamon Press.

Le Heron, R. (2003) 'Creating food futures: reflections on food governance issues in New Zealand's agri-food sector', *Journal of Rural Studies*, 19, 1: 111–125.

Lockie, S. (1999) 'The state, rural environments, and globalisation: "action at a distance" via the Australian Landcare Program', *Environment and Planning A*, 31: 597–611.

McMichael, P. (ed.) (1994) *The Global Restructuring of Agro-Food Systems*, Ithaca, NY: Cornell University Press.

McMichael, P. (2000) *Development and Social Change: A Global Perspective* (2nd edn), Thousand Oaks, CA: Pine Forge Press.

Marsden, T., Lowe, P. and Whatmore, S. (eds) (1990) *Rural Restructuring: Global Processes and their Responses*, London: Fulton.

Marsh, S.P. and Pannell, D.J. (2000) 'Agricultural extension policy in Australia: the good, the bad and the misguided', *The Australian Journal of Agricultural and Resource Economics*, 44, 4: 605–627.

Martin, P. (1997) 'The constitution of power in Landcare: a post-structuralist perspective with modernist undertones', in S. Lockie and F. Vanclay (eds) *Critical Landcare*, Wagga Wagga: Centre for Rural Social Research, Charles Sturt University.

Michael, M. (2000) *Reconnecting Culture, Technology and Nature: From Society to Heterogeneity*, London: Routledge.

Miller, P. and Rose, N. (1990) 'Governing economic life', *Economy and Society*, 19, 1: 1–31.

Murdoch, J. and Ward, N. (1997) 'Governmentality and territoriality: the statistical manufacture of Britain's "national farm"', *Political Geography*, 16, 4: 307–324.

O'Malley, P. (1996) 'Indigenous governance', *Economy and Society*, 25, 3; 310–326.

Pretty, J. and Chambers, R. (1994) 'Towards a new learning paradigm: new professionalism and institutions for a sustainable agriculture', in I. Scoones and J. Thompson (eds) *Beyond Farmer First: Rural People's Knowledge, Agricultural Research and Extension Practice*, London: Intermediate Technology Publications.

Rose, N. (1993) 'Government, authority and expertise in advanced liberalism', *Economy and Society*, 22, 3: 283–299.

Rose, N. (1999) *Powers of Freedom: Reframing Political Thought*, Cambridge: Cambridge University Press.

Rose, N. and Miller, P. (1992) 'Political power beyond the state: problematics of government', *British Journal of Sociology*, 43, 2: 173–205.

Star, S.L. (1991) 'Power, technology and the phenomenology of conventions: on being allergic to onions', in J. Law (ed.) *A Sociology of Monsters: Essays on Power, Technology and Domination*, London: Routledge.

Ward, N. and McNicholas, K. (1998) 'Reconfiguring rural development in the UK: Objective 5b and the new rural governance', *Journal of Rural Studies*, 14, 1: 27–39.

Winter, M. (1997) 'New policies and new skills: agricultural change and technology transfer', *Sociologica Ruralis*, 37, 3: 363–381.

Woolgar, S. (1991) 'Configuring the user: the case of usability trials', in J. Law (ed.) *A Sociology of Monsters: Essays on Power, Technology and Domination*, London: Routledge.

Part III

(Re)configuring objects and subjects of governing

9 Sustainability and agri-environmental governance

Jacqui Dibden and Chris Cocklin

Introduction

The sustainability discourse has been remarkably persistent, indeed growing, in its political and social stature over the twenty years or so since it first rose to prominence. In agriculture the already well-established interest in sustainability has been fuelled recently by pressures to extend the use of bio-technology, an increasing consumer interest in food safety, a global focus on food security, and emerging public, political and consumer demands for a 'clean and green' food supply. Despite the social and political interest in sustainability, the 'agri-industrial' model of agriculture remains persistent. Economic and political agendas continue to promote a globalized food system, mass food markets, competition and efficiency, along with what Jessop (1997) refers to as the 'hollowing out' of the state. According to Marsden (2003), powerful interests, supported by regulatory and social networks, have been successful in upholding an inherently unsustainable accumulation system, but the 'crisis' of the extant accumulation system has deepened in the face of the sustainability agenda. Thus agricultural policies that seek to entrench the agri-industrial model increasingly stand in contrast to regulation and governing in respect of the environment. For farmers, this presents the dilemma of at once responding to supposedly unfettered markets while simultaneously satisfying increased regulatory and social expectations in terms of their environmental performance and other requirements (e.g. in relation to food quality). How these respective influences are mediated at the farm level in terms of decisions about land use and resource allocation is a question of some importance.

In this chapter we reflect on the environmental and sustainability policy agendas in relation to agriculture, drawing on a case study of the dairy industry in Australia. In the first section we outline key elements of contemporary Australian agricultural policy. We then consider aspects of agri-environmental governing. The discussion moves to the dairy case and, against the background of recent price deregulation, we map out the increasingly complex regulatory space in which Australian dairy farmers operate. By taking the broad theme of governance as a platform for discussing the regulation of agriculture, we aim to acknowledge the 'co-implications' (Castree 2002) of the social, economic and environmental. Within this context we raise the question of the sustainability of a more 'multifunctional' countryside.

'Competitive productivism' and the neo-liberal state

In agriculture and natural resource management there has been a shift from govern*ment* to new forms of *governance*, involving not only state agencies but also a range of other organizations from both private and public sectors. On these new forms of governing, Goodwin (1998: 10) remarks that the governance perspective leads 'us to look afresh at the old distinction between market, state and civil society'. According to Goodwin (1998: 9), 'The role for government is seen as one of identifying stakeholders and then developing the relevant opportunities and linkages for them to be brought together to act for themselves.' Similarly, Little (2001: 98) refers to 'a new role for the state – in particular its retreat from a welfarist position as provider of support to one of co-ordinator and manager of the various participants in the process of governance.'

In Australia, the changing role of government was associated from the 1970s with the replacement of state-sponsored and subsidized productivism by a new form of highly productive agriculture, which might be called 'competitive' or 'globalized' productivism – i.e. productivism shaped by neo-liberalism. The policy commitment to competitive productivism was expressed through opening up agriculture to competition on the world market, combined with the promotion of high-tech, intensive farming practices, farm amalgamations, training to increase 'capacity', and other efficiency measures (Argent 2002; Gray and Lawrence 2001; Higgins and Lockie 2002).

The realignment of political and economic actors is also evident in the 'deregulation' of economic activities, which has found particularly strong representation in the neo-liberalist restructuring that has been pursued vigorously by countries such as New Zealand and Australia. Deregulation refers to changes in the policies, rules and regulations instituted and enforced by the state, characterized by greater reliance on market forces and less government intervention. These regulatory changes have been justified with reference to the central tenets of the neo-liberalist agenda, namely economic efficiency, transparency and accountability, and an assumed superiority of market competition over government involvement; globalization features by virtue of its deployment as a rationale for deregulation and for dismantling supports for agriculture.

Globalization and environmental governance

The small size of the Australian domestic market, compared with the European Union and the USA, and the consequent importance of exports to the national economy, makes Australian primary production peculiarly susceptible to the demands of overseas markets. This has a number of implications for agri-environmental governing. First, in an attempt to overcome the trade barriers (import quotas and subsidies) erected by major trading partners, Australia has become a strong advocate of free trade. The arguments in favour of dairy deregulation exemplify this discursive position. A dilemma for Australian governments, however, is the perceived incompatibility of providing financial support for farmers, to help them deal with environmental problems arising from farming activities, with the view held by Australia and other Cairns Group nations that this would constitute a thinly disguised 'non-tariff barrier', and therefore be contrary to World Trade Organization (WTO) rules (Potter and Burney 2002). This view also informs the Australian government rejection of 'multifunctionality' – the idea that agriculture has environmental and social as well as economic functions – which has become a cornerstone of EU agricultural and trade policy (Hollander 2004; Potter and Burney 2002). This stance may be seen as one of the underlying motivations for the Australian government's reluctance to provide financial rewards (apart from small, short-term incentives) to landholders for environmental management.

Second, Australia has sought increasingly to enter into trade agreements in the hope of acquiring more favourable access for its goods, including agricultural products, but these agreements may undermine existing environmental regulations, such as the ability of Australian governments to regulate access to water or to maintain strong quarantine protections against pests and diseases (Australian Conservation Foundation 2003). As McCarthy (2004: 330) points out: 'Environmental laws, regulations, and requirements can potentially be classified as illegal "barriers to trade" under many recent trade agreements.' The trade agreements may therefore contradict another aspect of Australian government policy – the close attention it advocates to meeting the requirements of consumers and retailers in major overseas markets, including any actual or anticipated environmental standards. As a recent report argued: 'Australian standards must be continually evaluated for their equivalence to changing European standards, and should be improved to *meet or exceed European standards* where necessary' (Agriculture, Fisheries and Forestry Australia 2002: 3; emphasis added).

Contrasting sharply with the Australian government's sensitivity to overseas markets is the approach that has been taken under the present Liberal–National Coalition (conservative) government to international environmental agendas. As one critic (Christoff 2002: 51) observed:

> Overall – and in stark contrast to its stance on trade treaties and the workings of the World Trade Organization (WTO) – Australia has pursued and promoted voluntary measures in its environmental treaty negotiations and often been a stern opponent of mandatory measures, binding targets and related compliance mechanisms.

On the domestic front, the Federal government has also shown a preference for voluntary over regulatory measures in relation to the environment. Although the ability of the government to implement regulatory tools in support of environmental policy is well established, under recent federal legislation (the Environmental Protection and Biodiversity Conservation Act 1999) this responsibility has, in practical terms, been largely devolved to the states. However, the regulatory regimes established by the states have been influenced strongly by the Federal government's role under the Act to approve state-level procedures (known as 'bilateral agreements') and through the government's financial powers.[1] In terms of the latter, a potent means of control in recent years has been the National Competition Policy (NCP), which provides payments to the states on the basis of their progress in achieving competition reforms.[2] The threat by the Federal government to withhold these payments has been used to encourage compliance with a number of measures affecting agriculture, notably the (increasingly contested) deregulation of agricultural sectors and the introduction of water reforms. The latter have involved the establishment of transferable water rights that are separated from land rights and may be traded with other landholders (Productivity Commission 1999).

The preferred model of the Federal government for improving environmental management has been the participatory, self-help approach epitomized by the National Landcare Programme. Government promotion and support for local self-help has been viewed by several commentators as part of the neo-liberal shift from state to private regulation or, in this case, to 'self regulation of individuals' (Martin and Halpin 1998: 448; see also Higgins *et al.* 2001). Landcare forms part of a suite of natural resource management (NRM) programmes which 'seek to shape farmers' conduct in economically "rational" ways' (Higgins and Lockie 2001: 102–103). Other examples are the former Rural Adjustment Scheme, Property Management Planning, and the National Drought Policy. The latter has been 'designed to "reward" the good manager while offering exit support to those without a long-term sustainable future in farming' (Botterill 2003a: 67). However, this emphasis on good business management may be at odds with environmental responsibility, particularly where landholders' actions contribute to the public good rather than providing private benefits. As Higgins and Lockie (2001: 103) point out:

> The effect [of NRM programmes] is a rather restricted focus on environmental responsibility and sustainability as individual economic problems that can be addressed solely through improved business management. This has the tendency to equate profitability with environmental sustainability, thereby categorizing less profitable landholders as poor managers of natural resources and further legitimating programmes that seek to facilitate the exit of these producers.

The equation of profitability with environmental sustainability, and the assumption that farmers should bear the risks associated with farming, have weakened the effectiveness of Landcare as a mechanism for environmental improvement, since farmer participants – acting in accordance with sound business practice – have shown a marked preference for measures leading to immediate on-farm productivity benefits (e.g. perennial pastures, trees to shade livestock) rather than beneficial off-site effects (e.g. revegetation of saline ground water recharge zones) (Higgins and Lockie 2001). However, there is a growing body of evidence questioning the assumption that inculcating an 'environmental ethic' will produce more sustainable land management practices, particularly if these must be undertaken by farmers at their own expense. An evaluation of the National Landcare Program (Dames and Moore 1999: 73) revealed:

> While many landowners may be aware and committed to sustainable natural resource management practices, they may not have the financial resources to adopt these, even though they know that not adopting them may be to their own peril in future.

Hence the greater awareness generated by Landcare of the extent of local-level environmental damage has been seen by Martin and Halpin (1998: 455) as leading to a 'realization that substantial financial assistance is needed for ameliorating degraded resources, especially under the variable economic conditions brought about by economic deregulation'.

In 1999, at the end of the National Decade for Landcare (launched in 1989), the Federal government produced a national discussion paper, *Managing Natural Resources* (National Natural Resource Management Task Force (NNRMTF) 1999). This made explicit a policy shift away from reliance on voluntary, local community-based activities towards community–government 'partnerships' and an emphasis on regions (rather than local areas). Individuals still had a role to play as 'leaders and "champions" in the community who can demonstrate benefits and motivate actions for sound natural resource management' (NNRMTF 1999: vii). However, the paper also proposed facilitating fundamental changes in resource use and NRM through the employment of economic instruments, including taxation and other incentives, and through regulation and quality-assurance accreditation schemes (NNRMTF 1999: 48–53). The NCP water reforms were seen as 'an example of how property rights and market incentives can be used to achieve efficient and sustainable management of resources' (NNRMTF 1999: 48). Other proposed incentives included stewardship payments 'providing an income to landholders who manage their land for conservation and provide wider environmental and social "products" for natural resource management purposes' (NNRMTF 1999: 41).

With the development of the National Action Plan for Salinity and Water Quality in 2001, the Federal government moved towards agreements with the states as 'joint investment partners in natural resource management activities', according to which regional-level Catchment Management Authorities (CMAs) are charged with the task of implementing management plans (Commonwealth of Australia 2001). This new structure of governing follows the neo-liberal pattern of devolution of responsibility (in this case to the regions), expressions of support for community involvement and 'partnership' arrangements, and an emphasis on identifying and achieving 'objectives ... targets, activities, milestones, performance measures at the regional level, [and] a comprehensive monitoring and evaluation process' (Commonwealth of Australia 2001). Targets are to be achieved through land and water reforms, involvement of community groups, building the capacity of communities and landholders, and strategic 'investment' in CMA strategies and activities. The emphasis on targets and monitoring may be a response both to the criticism of Landcare for failing to evaluate environmental outcomes to ensure that funds were well spent, and also to bureaucratic requirements for 'accountability'. Although much of the change is to be accomplished through persuasion, education, information and incentives, some coercive power is provided by state-level legislation regulating water use and protecting native vegetation.

In addition to this new structure of governing, there has also been a shift – similar to that reported for New Zealand – towards extra-state regulation. Le Heron and Roche (1999: 204) argue that agricultural restructuring and deregulation in New Zealand have not removed regulation: 'Instead, agriculture has been reregulated in a potentially more complex way ... through greater market disciplines, with an emphasis on certain kinds of *political* intervention ... and is not simply a rolling back of the state.' In Australia, the state has sought to intervene in the market in support of national objectives. First, as we have shown above, the Federal government has experimented with the use of market-based instruments, such as tradable water rights. These are seen as a means of ensuring that water will be used more efficiently. There has also been an attempt to make use of the governing capacities of the actors in global commodity chains, through promotion of environmental management systems (EMS) and 'production accreditation' as means to address the concerns of both domestic and international consumers 'about the environmental and ethical impacts of production systems' (NNRMTF 1999: 50).

Adoption of 'voluntary' codes of practice and accreditation (such as ISO 14000 certification) is advocated to counteract the risk that:

> if Australian industry does not ... establish internationally recognized standards for production systems such standards will be developed and imposed by external markets in response to consumer concern. *This could present trade barriers for our agricultural products in the future.*
>
> (NNRMTF 1999: 50; emphasis added)

Recognizing the reluctance of farmers to be regulated, Federal government representatives have been at pains to emphasize the voluntary nature of EMS, as a speech to the Victorian Farmers Federation by the Parliamentary Secretary for Agriculture, Fisheries and Forestry (Troeth 2003) reveals:

> It is absolutely essential that industry remain in the driving seat of EMS. Let's not make any mistake about that. There have been some views that this is one way for governments to legislate farmers, to govern practices at a farm level. Let me categorically set the record straight ... government's role in EMS is to support industry, not shackle it – to facilitate both NRM outcomes, production efficiencies, and ensure market access.

However, to underline the dangers of *not* adopting EMS, Senator Troeth recounted her experiences on a recent visit to Europe:

> I think it is fair to say that, in contrast to the Commonwealth government's view, many Europeans, and the EU in general, are becoming far more prescriptive in the regulation of farming systems and in particular where we are concerned – what they require of imports into the EU.

Summing up, the environmental policies and programmes of the Commonwealth government are characterized by a preference for market-based solutions, devolution to lower scales (including the individual landholder), and what we call *'rhetorical action'* – action taken with the intention of persuading or influencing, rather than with an expectation of producing results in its own right. This is not to suggest that funding for environmental action has not provided a useful resource for those landholders and community groups who have been able to obtain it. However, the amount of funding provided is not – in itself – sufficient to make a substantial contribution to reversing environmental deterioration (Christoff 2002). Moreover, it is to a large extent undermined by policies promoting more competitive, productive (and therefore often intensive) agriculture, which tends to be associated with detrimental environmental impacts. Farming has increasingly been viewed by policy-makers primarily as a business (operated by 'farm business managers') rather than as a way of life, a change in attitude supported by a shift in rhetoric (Botterill 2003b: 201).

The emphasis in government discourse on landholders becoming self-sufficient, competitive business managers is at odds with the expectation that farmers and other rural people will undertake voluntary environmental work for the public good and not primarily for private benefit. This dissonance has been seen by Buttel (2003: 26) as a broader feature of government policy in Western nations:

> contemporary nation-states intrinsically have very contradictory imperatives relating to environmental quality and resource consumption. . . .

Thus, instead of assuming that the state plays a relatively unitary role regarding the environment, it is more useful to anticipate that state policy will be contradictory and that conflicts over environmental protection and resource conservation will be played out within the state itself.

The persistent lack of policy integration in Australia has been remarked on by Toyne and Farley (2000: vii) in a review of *The Decade of Landcare*:

A clear deficiency with the policy environment of Landcare has been the failure to properly articulate its place in the bigger picture. Structural adjustment, market systems, macroeconomic policy and economic incentives are all disconnected from Landcare policy. So too are issues such as State government responsibilities, regional structures, service provision and incentives.

The current enthusiasm for EMS may be seen as an attempt to reconcile the competing elements of Federal government policies for agricultural production and the environment, by associating improved environmental practices with the possibility of enhanced market access and by attempting to enlist supply-chain participants in this endeavour. In the case study of dairying which follows, we trace the changing structures of regulation and governing produced both by neo-liberal economic reforms and environmental initiatives in dairy regions, particularly in Victoria – the state predicted to be most favourably affected by deregulation (Cocklin and Dibden 2002a, 2002b).

Dairy deregulation

The Australian dairy industry provides one recent example of how the tenets of neo-liberalism were drawn upon to construct a rationale for deregulation (Cocklin and Dibden 2002a, 2002b). In spite of the sector's global exposure, Australian dairy producers were, for a long time, protected under statutory marketing arrangements. Post-farm-gate prices were deregulated in all states from 1990, but farm-gate prices remained regulated. During the late 1990s this was subject to scrutiny on the grounds that it was contrary to National Competition Policy (NCP) and discouraged innovation and efficiency (Productivity Commission 1999). It was also argued that the extant regulatory arrangements weakened Australia's position in the WTO negotiations for trade liberalization and for dairy industry reform by other countries (Senate 1999). Collectively, the arguments in support of regulatory change centred on issues of international competitiveness and efficiencies in production, forming part of what Pritchard (1999: 421) referred to as 'Australia's neo-liberal project to restructure national institutions in line with market principles'.

The deregulation of farm-gate prices came into effect on 1 July 2000, and was accompanied by the removal of all price supports and the dismantling of state statutory marketing authorities (Cocklin and Dibden 2002a, 2002b). A financial compensation package could be used by farmers to facilitate exit from the industry or to ameliorate the impact of deregulation through the retirement of debt or investment in new capital, land and so on.

The economic consequences of deregulation were very much as predicted. The immediate effects were 'milk wars', which drove down prices paid by supermarket consumers and hence farm-gate prices for milk. In all states except Victoria, there were reports of large numbers of dairy farmers leaving the industry (ABARE 2001; Davidson 2001). In Victoria, by contrast, there were few adverse effects in the short run, due to sharp rises in export prices – which coincided fortuitously with deregulation – and the high degree of previous farm rationalization (Cocklin and Dibden 2002b). In subsequent years, however, unfavourable exchange rates have reduced both export and linked domestic prices, while a prolonged and severe drought has increased farmers' feed costs and reduced access to water for irrigation. Dairy farm numbers in Victoria fell by almost 20 per cent in four years from 7,800 in 2000 (before deregulation) to 6,200 in May 2004, with nearly one-third leaving in the previous eight months (Jackson 2004). Their departure was attributed to 'the advancing age of the average dairy farmer and a lack of young people entering the industry', but also to 'the tight margins farmers are getting. They are not being paid enough for their milk' (Jackson 2003: 18).

Consolidation of dairy farms and increased production were anticipated outcomes of deregulation and have been identified as an appropriate response to unfavourable terms of trade:

> In response to declining terms of trade, dairy farmers can increase the size of their operations (larger farms, more cows) and the intensity of their operations (higher stocking rates, more intensive production practices). Australian farmers have done both. The size of the average dairy farm has increased by more than 40 per cent in the past 15 years. The number of cows milked per farm has increased by nearly 75 per cent, while milk yield per cow has risen by more than 40 per cent. The net result has been a 160 per cent increase in milk production per farm in 15 years.
>
> (ABARE 2001: 3)

Increases in profitability have not necessarily followed, however.

Dairy restructuring and environmental crisis

The commitment to the tenets of neo-liberalism sits in uneasy juxtaposition with a growing recognition of environmental vulnerability. Salinity, soil erosion and prolonged periods of drought have been major elements of an environmental crisis that has exacerbated the social and economic problems confronting Australian farmers and rural communities (Gray and Lawrence 2001).

Remarkably, in the National Competition Policy reviews conducted by each state in the lead-up to dairy deregulation, the potential environmental impacts of deregulation were largely disregarded, despite the fact that the Net Community Benefit Test under the National Competition Policy requires that 'government legislation and policies relating to ecologically sustainable development' should be taken into account (Senate 1999: 66). In one of the few references to environmental outcomes in a report on the proposed deregulation of the Australian dairying industry, a Senate Committee inquiry observed, on the basis of public submissions and other evidence, that:

> Less profitable farms will mean working resources to their limits, including:
>
> a   Running larger herds on existing properties;
> b   Intensive stocking, feeding and milk production resulting in increased waste disposal problems;
> c   Maximum use of fertilizers, etc to improve pasture productivity; and
> d   Increased demands on limited water resources in pasture irrigation to increase productivity.
>
> (Senate 1999: 134)

The Senate Committee (1999: 121) drew a link between environmental management and predictions that dairy farms would become larger under deregulation, noting from submissions 'that the ultimate impact of larger farms will be . . . less investment in the maintenance of the environment and development of environmental approaches to minimize the impact of dairying on the community'.

In the lead-up to deregulation, a survey carried out for the National Land and Water Resources Audit also pointed to the prospective tension between the facilitation of the agri-industrial model and pressures for a more ecologically sustainable agriculture (Day 2000: 14):

> Deregulation – and resulting lower prices and increased uncertainty in some areas of the dairy industry – may reduce the capacity to invest in environmental management at a time when the same pressures are promoting more intensive production, which necessitates higher levels of environmental management.

The structures of governing for dairying, which have evolved over the past few years in Australia, results from the conjunction of dairy price deregulation with the reregulation of the environment. Dairy farmers and other rural producers have been subjected to a range of potentially contradictory global and national forces leading to deregulation in some arenas (e.g. removal of trade barriers, price deregulation), coincident with increased regulation in respect both of the environment (Productivity Commission 1999: xxvii) and food quality. The tension between efforts to maintain an internationally competitive dairy industry and to achieve environmental sustainability has been particularly evident in recent years, when dairy farmers have confronted both the worst drought for twenty years (Drought Review Panel 2004) and the trials of competing without government support on a far from open world market. Against the background of the deregulation of dairying, we look next at the changing regulatory structures, drawing on three specific examples: water resource policy, quality assurance, and the implementation of environmental management systems (EMS).

Water resource management

The situation of dairy farmers in irrigation regions has been exacerbated by the juxtaposition of two reforms introduced under National Competition Policy – deregulation of milk marketing and reforms to water regulation, in particular the introduction of tradable water rights. Victoria has a relatively well-established and active water market in irrigation areas,[3] with water able to be traded out of state water authority districts and even interstate up to a limit of 2 per cent per annum (Department of Sustainability and Environment (Victoria) 2003: 68). The ability to sell water at a time when dairying is experiencing a severe cost-price squeeze has resulted in the migration of water rights from northern dairying districts to the higher value wine and horticultural industries of Sunraysia in north-western Victoria and to South Australia (Department of Sustainability and Environment (Victoria) 2003: 69). Cumulative shifts in water rights can lead to potential loss of water access for a particular locality or district, since the costs of maintaining distribution channels fall on to a dwindling pool of irrigators. The answer envisaged by the Victorian government is to provide assistance for farmers to 'adjust' to closure of the irrigation system for 'areas not suitable for irrigation – and where we must encourage people not to irrigate' (Hunt 2003).

Another threat to irrigators' water security are the proposals at both Federal and state levels to return more water to rivers as environmental flows. Proposals to reinstate environmental flows, along with other water management strategies, illustrate graphically the divergence between the expansionist, businesslike behaviour that farmers have been encouraged to adopt and the regulatory restrictions now proposed on behalf of the environment. The effect of managing with less water will be that the dairy industry will have to either source 'other water . . . or fodder to maintain production *or reduce production*' (Murray Dairy 2003: 14; emphasis added). Substantial declines in production and incomes are forecast, but also decreased ability of farmers to make environmental improvements.

Water reforms form a component of a broader structural adjustment pro-gramme. The assumption underlying this policy is that the introduction of property rights in water, which are separated from landownership, will enable resources to be redistributed to the most profitable farm businesses. However, a study by Bjornlund (2003: 142) showed that those farmers who were engaged in structural adjustment had incurred debts in order to expand their land and water rights, with potentially serious consequences.

Food safety, quality and the environment

Deregulation in mid-2000 resulted in the dismantling of the previous state dairy authorities and their replacement in all states with food safety bodies. In Victoria, unlike the other states, there is a dedicated dairy organization, Dairy Food Safety Victoria (DFSV), which works in 'partnership' with industry. DFSV licenses and audits processing factories and vendors, as well as monitor-ing 'programmes within factories to ensure they audit their farmers' (*Dairy Page* 2000). To meet this obligation, licensed dairy processors (co-operatives and companies) have been introducing quality assurance programmes for their farmer-suppliers. For example, the Murray Goulburn co-operative – the largest processor in Victoria – has been active in promoting 'Milkcare', a pro-gramme that requires farmers to maintain high (and clearly defined) standards of food safety, but not – as yet – specified environmental standards. However, there is increasing concern that environmental certification may be imposed by overseas buyers (Agriculture, Fisheries and Forestry Australia 2002). The state food safety bodies are currently engaged in developing 'a national regulatory framework for the dairy industry' (Farmonline 2003), thus underlining the fact that *de*regulation (so-called) is in fact *re*regulation – a shift from one form of regulatory structure to another. Indeed, Marsden (2003: 169) argues that:

> The agro-industrial logic [of food production] has been supplanted in some ways by a parallel more bureaucratic mode of regulation and perspective. This has, in highly sophisticated ways, attempted to 'correct' the agro-industrial model rather than construct viable alternat-ives to it. It has done this ... by attempting to put into place, in a highly interventionary and bureaucratic fashion, policies which effect-ively 'police' the food and rural system in ways which seem to make them more 'hygienic' and environmentally safe spaces of production and consumption.

Cocklin and Dibden (2002b: 39) observed, similarly, that:

> Stricter requirements in terms of quality assurance have led some farmers to feel that one form of regulation – which benefited them

financially – has been replaced by another form which limits the auto-nomy which made the physical hardships of dairying worthwhile; it has also inflated the debt burden of some farmers.

Whereas quality in the European literature has often been associated with local or speciality produce (e.g. Murdoch *et al.* 2000), in Australia it consti-tutes a way of differentiating and giving an edge to bulk commodities in world markets; for example, by linking Australian foodstuffs with Aus-tralia's 'clean and green' image (Lawrence 2005) or by demonstrating that these foods are 'safe'. Both these motivations may provide an explanation for opposition by two dairy processors (Tatura Milk and Murray Goulburn Co-operative) to the introduction of genetically modified (GM) crops, which could potentially contaminate feed for dairy cattle, and damage the 'GM-free' image of Australian dairy produce. These processors have been success-ful in persuading the Victorian government to introduce a four-year moratorium on commercial cultivation of GM crops (Farmonline 2004a).

Governing farmer behaviour – EMS and Landcare

Higgins and co-workers (2001: 212–221) regard Landcare as an example of 'rationalities and techniques through which governments and their agencies exert some form of "action at a distance" in Australian agricultural environ-ments', persuading and improving the capacity of farmers to regulate their own behaviour as good managers.

> In this way, individual preferences and choices are aligned with govern-mental objectives through the enhancing of self-regulatory capacities, rather than by means of coercion.
>
> (Higgins *et al.* 2001: 215)

Similarly, environmental management systems (EMS) – which the federal government is promoting both through regional bodies and farming sectors – may be seen as a technology of governing farmer behaviour through sub-national structures and the commodity chain, by involving them in the development of codes of practice and regional environmental standards. The adoption of EMS is currently being stimulated through funding for pilot projects under the 'Pathways to Industry Environmental Management Systems (EMS) Programme' (Farmonline 2004b). This initiative reveals the difficulty governments appear to experience in providing *environmental* justi-fications for programmes with environmental objectives. While funding comes from the Natural Heritage Trust – a major source of environmental funding in Australia – the advantages of EMS are stated primarily in *economic* terms by the government:

> As well as looking to boost profitability and sustainability, it will also

> help our farmers better demonstrate their environmental stewardship to
> the community *and to overseas markets.*
>
> (Farmonline 2004b; emphasis added)

The preoccupation with demonstrating good environmental credentials to
the outside world has been seen by Sonnenfeld and Mol (2002: 1326) as a
feature of globalization:

> environmental concerns and pressures may arise in one corner of the
> globe and rapidly be generalized around the world through combined
> force of market, media/cultural, regulatory and political actions.

Overseas markets also provide a way of 'selling' the idea of EMS – as a means
of supporting claims to food 'quality' and 'regional' characteristics – to
farmers and farming bodies. However, other motivations, such as awareness
of adverse public opinion, may be equally potent. For example, a dairy
environmental self-assessment tool was developed 'at the request of several
Victorian dairy farmers . . . asking for an environmental tool they could use
to prove to the wider community that they are doing a good job of caring
for their farms and the environment' (Dairying for Tomorrow, n.d.).
Funding was subsequently provided by the Commonwealth EMS project.

While the notion of governmentality has considerable explanatory power,
it should not be assumed that only *state* agencies are engaged in attempting
to shape the behaviour of rural producers. Farmer organizations, industry
bodies and many other actors within the supply chain appear equally committed to the discourse of neo-liberalism. Moreover, farmers and other rural
people are not passive recipients of influences from above. They are often
quite skilled at extracting from government-funded programmes what they
want and rejecting the rest, forcing governments to modify their offerings.
An example of this is the poor uptake of federal grants offered for individual
farmers to undertake EMS until changes were made to improve the incentives offered and to make the conditions less onerous.

This quiet resistance to government direction is indicated in the response
of farmers to the Landcare programme. Apart from the fact that over 60 per
cent of landholders Australia-wide have not joined Landcare groups (Nelson
et al. 2004: 15), those who have show variable degrees of receptivity to the
activities eligible for funding in accordance with programme priorities.

Conclusions: reregulation and post-productivist tendencies

As our dairy case study illustrates, there is an increasingly apparent incompatibility between deregulated, competitive, intensive agriculture, and the
notion of rural sustainability. In particular there are concerns for the condition of rural communities and over the widening environmental crisis that

threatens the productivity of agriculture, as well as the health of rural towns and natural ecosystems (Marsden 2003). This environmental and social crisis has forced a reappraisal of policy directions grounded in neo-liberal understandings and discourse. In Australia, despite continuing to promote neo-liberal principles, 'a substantial shift has occurred in the area of agricultural and natural resource management policy' (Argent 2002: 105). There are indications that governments at both state and federal level increasingly recognize that farmers cannot be expected to bear the burden alone for 'public good' environmental work and neither can rural communities be expected to deal single-handedly with their decline. This contrasts with the willingness of governments in the EU and USA to provide funding for a range of agri-environmental schemes (see e.g. Potter 1998; Wilson and Hart 2001).

The question arises as to the extent to which Australian agricultural and environmental mechanisms of governing represent a movement away from productivism towards promotion of a multi-functional and sustainable countryside, and how Australian farmers are reacting to such policies. In this context, the nature and trajectory of policy-making will be influenced by dimensions such as prevailing ideologies of relevant stakeholders, the nature of changing food networks in increasingly globalized agro-commodity chains, shifts in agricultural techniques towards more environmentally friendly farming methods (which can be a response as well as a driver of policy change; cf. Ward 1993), and the reconfiguration of productivist/post-productivist actor spaces and forms of governing.

In the UK, the WTO negotiations have triggered a debate 'about how far and in which way to sustain a multifunctional agriculture and an agrarian policy agenda under increasingly liberalized market regimes' (Potter and Burney 2002: 46), while Australia is grappling with the opposite problem – how to combine an already liberalized economy with the need to move towards more sustainable land management and viable rural communities. The difficulties of reconciling these seemingly incompatible directions provide lessons for other countries contemplating a more 'liberal' rural future.

Notes

1 In Australia, only the Commonwealth government has the right to levy taxes, which are then redistributed to the states under a wide range of funding agreements and mechanisms.
2 The NCP reforms were initiated in 1995. Payments to the states have risen from AUS$396.2 million in 1997/1998 to an estimated maximum of AUS$802 million in 2005/2006 (Productivity Commission 2004: 30).
3 Trade in water commenced in 1991 (Department of Sustainability and Environment (Victoria) 2003: 59).

References

Agriculture, Fisheries and Forestry Australia (AFFA) (2002) *Safely on the Shelf: The Impact of Global Retailers' Safety and Quality Requirements on Australian Food Exporters,* Canberra: Commonwealth of Australia.

Argent, N. (2002) 'From pillar to post? In search of the post-productivist countryside in Australia', *Australian Geographer,* 33, 1: 97–114.

Australian Bureau of Agricultural and Resource Economics (ABARE) (2001) *The Australian Dairy Industry: Impact of an Open Market in Fluid Milk Supply,* ABARE Report to the Federal Minister for Agriculture, Fisheries and Forestry, Canberra: Australian Bureau of Agricultural and Resource Economics.

Australian Conservation Foundation (ACF) (2003) *ACF Position on Australia–United States Free Trade Agreement,* available at http://www.acfonline.org.au (accessed 21 May 2004).

Bjornlund, H. (2003) 'The socio-economic structure of irrigation communities – water markets and the structural adjustment process', *Rural Society,* 12, 2: 123–147.

Botterill, L. (2003a) 'Uncertain climate: the recent history of drought policy in Australia', *Australian Journal of Politics and History,* 49, 1: 61–74.

Botterill, L. (2003b) 'Beyond drought in Australia: the way forward', in L. Botterill and M. Fisher (eds) *Beyond Drought in Australia: People, Policy and Perspectives,* Collingwood: CSIRO Publishing.

Buttel, F. (2003) 'Environmental sociology and the sociology of natural resources: strategies for synthesis and cross-fertilisation', in G. Lawrence, V. Higgins and S. Lockie (eds) *Environment, Society and Natural Resource Management: Theoretical Perspectives from Australasia and the Americas,* Cheltenham: Edward Elgar.

Castree, N. (2002) 'Environmental issues: from policy to political economy', *Progress in Human Geography,* 26, 3: 357–365.

Christoff, P. (2002) *In Reverse,* Report prepared for the Australian Conservation Foundation, available at http://www.acfonline.org.au/docs/publications/rpt0027.pdf (accessed 10 May 2004).

Cocklin, C. and Dibden, J. (2002a) 'Deregulating the Australian dairy industry', in P. Holland, F. Stephenson and A. Wearing (eds) *2001, Geography – A Spatial Odyssey,* Proceedings of the Joint Institute of Australian Geographers/New Zealand Geographical Society Conference 2001, Hamilton, NZ: New Zealand Geographical Society.

Cocklin, C. and Dibden, J. (2002b) 'Taking stock: farmers' reflections on the deregulation of Australian dairying', *Australian Geographer,* 33, 1: 29–42.

Commonwealth of Australia (2001) *Agreement between the Commonwealth of Australia and the State of Victoria for the Implementation of the Intergovernmental Agreement on a National Action Plan for Salinity and Water Quality,* available at http://www.napswq.gov.au/publications/bilaterals/vic/introductory1–8.html (accessed 16 August 2004).

Dairying for Tomorrow (n.d.) 'Environmental "self assessment" leads to improved priority setting', *Snapshots,* available at http://www.dairyingfor tomorrow.com (accessed 23 March 2004).

Dairy Page (2000) 'A change of authority', *Dairy Page: Dairy Farming Online,* Vol. 2 (August), availale at http://www.dairypage.com.au (accessed 29 January 2001).

Dames and Moore-NRM (1999) *NHT Mid Term Review – NLP for Agriculture, Fish-*

eries and Forestry Australia, available at http://www.nht.gov.au (accessed 1 July 2004).

Davidson, A.P. (2001) 'Restructuring the dairy industry in New South Wales', *Rural Society,* 11, 1: 23–37.

Day, P. (2000) *Natural Resource Management on Australian Dairy Farms: A Survey of Australian Dairy Farmers,* Canberra: National Land and Water Resources Audit.

Department of Sustainability and Environment (DSE) (2003) *Securing our Water Future: A Green Paper for Discussion,* available at http://www.dse.vic.gov.au.

Drought Review Panel (2004) *Consultations on National Drought Policy: Preparing for the Future,* Canberra: Department of Agriculture, Fisheries and Forestry.

Farmonline (2003) 'New team to drive dairy industry forward', *Farmonline,* 7 October, available at http://www.farmonline.com.au (accessed 7 October 2003).

Farmonline (2004a) 'Dairyfarmers back Brack's Government GM decision', *Farmonline,* 21 May, available at http://www.farmonline.com.au (accessed 24 May 2004).

Farmonline (2004b) '$5.2m to boost agricultural environmental management', *Farmonline,* 26 April, available at http://www.farmonline.com.au (accessed 24 May 2004).

Goodwin, M. (1998) 'The governance of rural areas: some emerging research issues and agendas', *Journal of Rural Studies,* 14, 1: 5–12.

Gray, I. and Lawrence, G. (2001) *A Future for Regional Australia: Escaping Global Misfortune,* Cambridge: Cambridge University Press.

Higgins, V. and Lockie, S. (2001) 'Neo-liberalism and the governing of natural resource management in Australia', in J. Dibden, M. Fletcher and C. Cocklin (eds) *All Change! Gippsland Perspectives on Regional Australia in Transition,* Melbourne: Monash Regional Australia Project Occasional Papers 2001.

Higgins, V. and Lockie, S. (2002) 'Re-discovering the social: neo-liberalism and hybrid practices of governing in rural natural resource management', *Journal of Rural Studies,* 18, 4: 419–428.

Higgins, V., Lockie, S. and Lawrence, G. (2001) 'Governance, local knowledge and the adoption of sustainable farming practices', in G. Lawrence, V. Higgins and S. Lockie (eds) *Environment, Society and Natural Resource Management: Theoretical Perspectives from Australasia and the Americas,* Cheltenham: Edward Elgar.

Hollander, G. (2004) 'Agricultural trade liberalization, multifunctionality, and sugar in the south Florida landscape', *Geoforum,* 35, 3: 299–312.

Hunt, P. (2003) 'Let some areas go dry, says Thwaites', *The Weekly Times,* 3 September.

Jackson, M. (2003) 'Dairy farmers shut up shop', *The Weekly Times,* 29 October.

Jackson, M. (2004) 'Enough is enough, say dairy farmers', *The Weekly Times,* 19 May.

Jessop, R. (1997) 'Capitalism and its future: remarks on regulation, government and governance', *Review of International Political Economy,* 4: 561–581.

Lawrence, G. (2005) 'Globalisation, agricultural production systems and rural restructuring', in C. Cocklin and J. Dibden (eds) *Sustainability and Change in Rural Australia,* Sydney: UNSW Press.

Le Heron, R. and Roche, M. (1999) 'Rapid reregulation, agricultural restructuring and the reimaging of agriculture in New Zealand', *Rural Sociology,* 64, 2: 203–218.

Little, J. (2001) 'New rural governance?' *Progress in Human Geography,* 25, 1: 97–102.

McCarthy, J. (2004) 'Privatizing conditions of production: trade agreements as neoliberal environmental governance', *Geoforum,* 35, 3: 327–341.

Marsden, T. (2003) *The Condition of Rural Sustainability,* Assen, The Netherlands: Royal Van Gorcum.

Martin, P. and Halpin, D. (1998) 'Landcare as a politically relevant new social movement?', *Journal of Rural Studies,* 14, 4: 445–457.

Murdoch, J., Marsden, T. and Banks, J. (2000) 'Quality, nature and embeddedness: some theoretical considerations in the context of the food sector', *Economic Geography,* 76, 2: 107–125.

Murray Dairy (2003) *Securing our Water Future – Green Paper for Discussion: A Submission on the Strategic and Economic Importance of the Northern Victorian Irrigated Dairy Industry,* Kyabram, Victoria, November.

National Natural Resource Management Task Force (NNRMTF) (1999) *Managing Natural Resources in Rural Australia for a Sustainable Future: A Discussion Paper for Developing a National Policy,* Canberra: Department of Agriculture, Fisheries and Forestry Australia, available at http://www.napswq.gov.au/publications/nrm-discussion.html (accessed 23 May 2004).

Nelson, R., Alexander, F., Elliston, L. and Blias, A. (2004) *Natural Resource Management on Australian Farms,* Canberra: Department of Agriculture, Forestry and Fisheries.

Potter, C. (1998) *Against the Grain: Agri-environmental Reform in the United States and the European Union,* Wallingford: CAB International.

Potter, C. and Burney, J. (2002) 'Agricultural multifunctionality in the WTO – legitimate non-trade concern or disguised protectionism?', *Journal of Rural Studies,* 18, 1: 35–47.

Pritchard, B. (1999) 'National Competition Policy in action: the politics of agricultural deregulation and wine grape marketing in the Murrumbidgee Irrigation Area', *Rural Society,* 9, 2: 421–441.

Productivity Commission (1999) *Impact of Competition Policy Reforms on Rural and Regional Australia,* Canberra: Australian Government Publishing Service.

Productivity Commission (2004) *Review of National Competition Reforms: Discussion Draft,* Canberra: Productivity Commission.

Senate Rural and Regional Affairs and Transport References Committee (1999) *Deregulation of the Australian Dairy Industry,* Canberra: Department of the Senate.

Sonnenfeld, D.A. and Mol, A.P.J. (2002) 'Ecological modernization, governance and globalization: epilogue', *The American Behavioral Scientist,* 45, 9: 1456–1461.

Toyne, P. and Farley, R. (2000) *The Decade of Landcare: Looking Backward – Looking Forward,* Sydney: The Australia Institute, Discussion Paper 30.

Troeth, J. (2003) *Speech to the VFF EMS Seminar, Melbourne, Victoria, 18 June,* available at http://www.affa.gov.au/ministers/troeth/speeches/2003/vffems2003.html (accessed 3 May 2004).

Ward, N. (1993) 'The agricultural treadmill and the rural environment in the post-productivist era', *Sociologia Ruralis,* 33: 348–364.

Wilson, G.A. and Hart, K. (2001) 'Farmer participation in agri-environmental schemes: towards conservation-oriented thinking?', *Sociologia Ruralis,* 41, 2: 254–274.

10 Governing consumption

Mobilizing 'the consumer' within genetically modified and organic food networks

Stewart Lockie and Nell Salem

Introduction

The history of food production, distribution and regulation is replete with attempts to influence and control the consumption behaviour of others. Examples range from direct state intervention in what substances are ingested (e.g. the iodization of salt) to more subtle attempts to influence the decisions people are likely to make through advertising, education and scientific advice (e.g. the 'Food Pyramid'). Attempts to influence food consumption are often characterized by conflict and debate. Attempts to mobilize people as consumers of genetically modified (GM) and organic foods in particular show this conflict to manifest in a range of social spheres including, broadly, regulatory processes and institutions, the media and public relations, social movement organizations and the marketplace (Babninard and Josling 2001). In this chapter we are not concerned with debates between proponents of GM and organic foods as such, but with the broader range of strategies deployed by proponents of GM and organic foods to mobilize people as consumers of these foods.

Governmentality and hybrid networks

The theoretical framework for this chapter draws upon elements of both the governmentality perspective and work within the sociology of science and technology (SST), or actor-network theory. The concept of governmentality reflects Foucault's conceptualization of power as a phenomenon that may take many forms, at times concentrated, hierarchical and repressive; at times creative and dispersed (Hindess 1996). Government, for Foucault (1991), is not an institution but an activity concerned with the 'conduct of conduct'. Underlying this activity are rationalities of governing that render objects knowable and actionable; defining the boundaries of acceptable intervention and offering strategies for that intervention. Of particular interest to

Foucault and other scholars have been the various forms of neo-liberal political rationality of increasing importance since the Second World War. The
feature of neo-liberal rationalities of most relevance here is their redefinition
of the individual. Where liberal ideologies constructed the individual as an
independent actor over whom the state may legitimately exert little influence, neo-liberal ideologies reconstruct the individual as a behaviourally
'manipulable being' who may be counted on to respond rationally to changing environmental variables (Lemke 2001: 200). Through the promotion of
market relations, neo-liberals have sought to influence the environment
within which people make decisions (Miller and Rose 1990), and the ways
in which they are likely to understand and respond to that environment
(Burchell 1993). 'Market reform' may thus be described as a 'technology of
the self'; an indirect technique to lead and control individuals without
taking responsibility for them (Lemke 2001).

Marketing and advertising may also be described as 'technologies of the
self'. Just as neo-liberal state agencies encourage prudent and entrepreneurial
behaviour on the part of individuals, marketing and advertising encourage
individuals to redefine themselves as consumers of particular kinds of goods
and services. Assembling 'the consumer' in this manner is achieved by ren-
dering consumer choice 'intelligible' in terms of 'individualized psychological factors' that may be 'understood and engaged with in a calculated
manner' in order to create connections between 'the active choices of potential consumers and the qualities, pleasures and satisfactions represented in
the product' (Miller and Rose 1997: 30–31). From the perspective of SST,
this may be understood as a form of 'action at a distance' (Latour 1987). Psychological market research and similar techniques of calculation and classifi-
cation are deployed not simply to record and relay facts about existing
desires or anxieties but to open once unreachable objects to calculation,
action and instrumentalization (Miller and Rose 1997: 31).

All objects are conceptualized by SST in terms of their relationships with
other objects (Law 1999). This has two consequences. The first is that there
is no distinction made between macro- and micro-levels of analysis (Latour
1999). Networks do not represent something bigger than the individual
objects that comprise them but the connections between those objects. Networks 'are nets thrown over spaces' that 'retain only a few scattered elements
of those spaces', embracing 'surfaces without covering them' (Latour 1993:
118). What appear to be macro-level social phenomena are, in fact, attempts
to sum up 'interactions through various kinds of devices, inscriptions, forms
and formulae' into 'very local, very practical' and 'very tiny' loci (Latour
1999: 17) by competing 'centres of calculation' that seek to speak for, and
thereby enrol, other actors in networks (Law 1994).

The second implication of the emphasis on the relationships between
objects in SST is the principle of analytical symmetry. The network
perspective decentres both the self and the collective as the focus of strategic
intention. Agency and power are conceptualized as emergent and variable

outcomes of relationships within networks rather than as properties of individuals (Callon and Law 1995; Law 1991). There is thus no reason to assume that power and agency take a consistent form, but there is also no reason to assume that they stem solely from relationships among humans. The networks of the social comprise a diverse array of people and things; 'hybrid collectives' of humans, nature and machines. While not necessarily ascribing agency or power in the conventional sense to non-humans, neither are they ascribed to humans. The principle of analytical symmetry requires researchers to examine the unique ways in which agency and power are constituted through each unique network while making no a priori assumptions about to whom, or what, influence may be attributed (Callon and Law 1995).

Drawing upon the governmentality perspective, and insights from SST, we are concerned in this chapter with the 'technologies of the self' and other strategies that are used to enrol people in networks of commodity production and consumption involving genetically modified and organic foods. While we will not offer a detailed analysis of a single genetically modified or organic food network from the laboratory to the farm to the stomach, the principle of analytical symmetry does remind us that forming and stabilizing the hybrid collectives of GM and organic food networks depends on the enrolment of many participants besides consumers. Hence it is unlikely that the enrolment of consumers can proceed successfully without some consideration of the relative success of would-be network engineers in mobilizing a range of other participants, including, in the case of GM foods, novel organisms.

Unruly natures and the overlapping topography of GM and organic networks

In 2003, an estimated 68 million hectares of land worldwide was sown to GM crops, a nearly fortyfold increase over the area planted to GM crops in 1996 that represented some 25 per cent of the total cropped area (Pew Initiative on Food and Biotechnology 2004). Genetically modified foods are those produced using recombinant-DNA technology, or genetic engineering (GE), to transfer specified pieces of DNA from one organism to another (Norton 2001). The precision and efficiency attributed to GE by its proponents over traditional plant and animal breeding is held to offer numerous benefits in terms of: human and animal health; animal and plant resistance to disease, pests and environmental extremes; production efficiency; product quality; the development of novel products; environmental protection; and economic growth and competitiveness (Norton 2001). According to critics, the precision and promise of GE are overstated (Ho 1998). Release and ingestion of genetically modified organisms (GMOs), they claim, is wrought with uncertainty, and is likely to result in reduced biodiversity and the creation of new ecological imbalances with unknown implications for human and environmental health (Ho 1998). Socially, critics claim that the property rights regimes associated with GE create relationships of economic and technical dependency between farmers and the biotech sector that are likely to promote rationalization of farm numbers and the ensuing decline of rural communities (Brac de la Perriere and Seuret 2000). The enthusiasm for GE among governments and agribusiness corporations, despite these concerns, may be accounted for both by popular scientific support for the doctrine of genetic determinism (i.e. the belief that each characteristic of an organism is controlled by a single gene or gene sequence) (Ho 1998; Juanillo 2001) and the numerous opportunities for vertical and horizontal integration of supply chains afforded by control over genetic material (Clapp 2002). The GE industry is thus dominated by 'life science' companies that have typically emerged through the integration of agrochemical, pharmaceutical, veterinary and seed firms (Babninard and Josling 2001) in order to optimize opportunities for the commercialization of research and development across multiple arenas of profit-making.

The organic food sector has also grown rapidly for well over a decade. Market growth rates in developed countries commonly estimated at 20 to 40 per cent per annum (Sahota 2004) have attracted considerable interest in organics among major retailers and food processors (Burch *et al.* 2001). Despite this, the sector remains dominated by small farms and producer-controlled industry organizations, and the amount of land devoted to organic production is estimated to be somewhere around half that dedicated to GM crops. In 2003, more than 24 million hectares worldwide were managed for organic production and a further 10.7 million were used for wild harvesting of plants that were subsequently certified as organic (Yussefi 2004). Somewhat less than half the area managed for organic production was arable (suitable for cropping) due to the use of large parcels of land in Australia and Argentina for extensive livestock grazing in semi-arid environments (Yussefi 2004). Importantly, these figures relate only to land that has been certified as meeting strict organic production standards by an independent certification organization. Certified organic food and agriculture is defined partly by what it is not (i.e. it is farming without the use of unapproved synthetic fertilizers, chemicals, animal growth promotants or GMOs) and partly by what it is (i.e. attempting to farm sustainably, enhance natural ecological processes and maintain high standards of animal welfare) (Sligh and Christman 2003). While much land that is not certified organic is managed, nevertheless, in a manner more or less equivalent to organic production standards, produce from that land cannot legally be sold as organic in the main Western markets for organic food.

The stability and growth of GM and organic food networks are dependent ultimately on the ability of proponents to enrol an exceptionally diverse array of people, processes and things, ranging from DNA sequences to soil–plant–animal interactions, regulatory apparatuses, consumers and so on. Yet both GM and organic food production systems may be conceptualized as responses to the general 'reluctance' of non-human organisms and processes to be enrolled in this manner. Where industrialized food networks require order, regularity and control, nature is often disorderly, unpredictable and uncooperative (Goodman *et al.* 1987). GM networks confront the unruliness of nature by reducing it to what is believed to be its most basic units – the proteins that comprise gene sequences – and thence manipulating these to express desired traits within organisms. Such manipulation has failed, however, to deliver the level of precision and control that the term genetic 'engineering' would imply. Field applications of GE are riddled with challenges ranging from the rapid buildup of resistance to genetically engineered plant toxins among target insects, to unforseen impacts on non-target species, disappointing yield results, genetic contamination of non-GE crops and cross-species transfer of GE material (Levidow and Carr 2000). Organic networks confront the unruliness of nature by seeking to mimic, and thence enhance, 'natural' processes of nutrient and energy recycling within biologically diverse agro-ecosystems. This too is potentially problematic. Contrary to the understanding most people have of organic systems as 'chemical-free', a limited range of synthetic inputs are allowed to deal with intractable pests for which there are believed to be no other practical control measures. There are thus many organic farms for which, apart from working with different, often naturally derived, inputs, production processes differ little from their conventional counterparts (Sligh and Christman 2003).

Despite the many obvious differences between GM and organic foods, the mobilization of networks around each is linked in several important ways. The most obvious is the material competition for network members. Just as important, however, are the discourses and practices of binary opposition through which both GM and organic foods are conceptualized (Campbell 2004). While organic agriculture has long been promoted as an alternative to input-intensive 'conventional' farming, widespread opposition to GM foods has seen interest in organic food increase dramatically and its oppositional status considerably enhanced. The organic industry has achieved a privileged place among opponents to genetic engineering due to its ability to identify both: tangible negative consequences for producers and processors, in terms of lost income, should their products be contaminated with genetically modified material; and politically, technologically and financially viable alternatives to GM technology (Campbell 2004). The relative success of GM and organic foods has become, for both networks, as much about their ability to speak on behalf of consumers, farmers, organisms and ecosystems through discourses of environmental and public health as it is about their ability to manipulate genetic and ecological processes in the service of food production.

Regulating GM and organic production and distribution

The neo-liberal underpinning of governmental rationalities has had major implications for the foci and form of state regulation of food production, trade and consumption. The preference for employing 'technologies of the self' over more direct forms of regulation is evident in the restructuring of agricultural production in line with the principles of market rule (Higgins and Lockie 2001); the management of environmental externalities caused by agricultural production (Lockie 1999); and the strategies deployed to cope with extreme environmental conditions such as drought (Higgins 2001). Despite this, GM and organic foods have attracted a raft of regulations covering activities ranging from production to distribution including laboratory research, intellectual property protection, field trials, trade, food safety and product labelling (Black 1998). It is not our intention here to analyse comprehensively the entire regulatory terrain of either GM or organic foods but to focus more specifically on the ways in which regulation of these sectors is oriented towards the governing of consumption.

Newell (2001) argues that the regulation of genetic engineering and its applications may more adequately be described as regulation *for* industry than as regulation *of* industry. Put another way, it may be argued that regulation of the production and distribution of GM organisms and foods is directed primarily towards influencing, 'at a distance', the behaviour of potential consumers. Despite the 'mass of legal regulations, non-legal rules, codes, circulars, practice notes, international conventions and ethical codes' relevant to GE (Black 1998: 621), and the equally complex array of government agencies, advisory groups, professional bodies and industry associations responsible for producing and monitoring them (Newell 2002: 1), regulation of genetic engineering and GM foods is organized predominantly around the principle of 'risk management' (Clapp 2002; Levidow and Carr 2000; Newell 2002). Risk management is used to mediate the contradictory imperatives confronting contemporary governments due to their dual roles as protectors of the public interest on the one hand, and as promoters of industrial and economic development on the other (Newell 2002). While there is some international debate over how risk management should be operationalized, the concept of 'substantial equivalence' is promoted by the USA, OECD and others as a means, based on 'sound science', to establish whether products are likely to raise safety concerns by comparing their physical and chemical properties with those of their conventional counterparts. Where GM foods are determined to be 'substantially equivalent', they are deemed suitable for treatment in the same manner as conventional foods with regard to safety and subjected to little, if any, additional testing. 'Substantial equivalence' facilitates rapid approval of GM foods by shifting the onus of responsibility away from the biotechnology industry to establish the safety of GM foods and on to its opponents to establish their riskiness (Levidow and Carr 2000).

The risk management regime has been contested fiercely (Norton and Lawrence 1996). This is most evident in relation to the issues of how far back from the point of ingestion the risks of GM foods should be assessed (Levidow and Carr 2000); and of labelling (Babninard and Josling 2001). European governments influenced by the 'precautionary principle', for example, have extended risk assessment to cover the entire food production process and, from April 2004, have introduced stricter labelling of GM foods (FSA 2003). Both of these moves have been criticized by the USA and others as non-tariff barriers to trade that are not based on 'sound science' (Levidow and Carr 2000). Nevertheless, defining the problem of GM food regulation in terms of risk management and substantial equivalence renders these knowable and manipulable by expert regulatory authorities. In so doing, it obscures scientific uncertainty and the value-laden nature of decision-making over the scope and boundaries of risk assessment by identifying only those risks that can be accurately measured or precedented as relevant (Newell 2002). Questions regarding the purpose and social implications of genetic engineering are ruled out as legitimate concerns, and a stable and predictable regulatory environment is established for business investment and trade. When coupled with other forms of state support for the development of biotechnology industries, risk management-based regulation may be seen as a neo-liberal strategy oriented directly to the creation of markets for biotechnology products (Newell 2002). Indeed, based on the application of 'sound science' by bureaucratic-scientific regulatory agencies, this regulatory regime accords the public almost no role in decision-making beyond the choices they may make as consumers within these new markets (Black 1998; Levidow and Tait 1995; Newell 2002).

Regulation of the organic food industry is, by comparison, relatively uncontroversial. Yet it is also, in many respects, oriented to the governing of consumption. Certification procedures for organic food do not, on the whole, stipulate how such food should be produced or processed but, rather, which inputs and practices should be avoided if that food is to be sold as certified organic (Sligh and Christman 2003). The focus of organic regulation is on guaranteeing the organic status of commodities as they are exchanged – on ensuring that consumers get what they pay for. However, before it is concluded that the regulation of organics is regulation *of*, rather than *for*, industry, it is pertinent to keep in mind that organic certification bodies are generally funded and run by producers. They exist, in other words, to create markets through the establishment of relations of trust that extend beyond co-present interaction.

'Technologies of the self' and mobilization of 'the consumer'

A constant theme within the sociological literature of the past two decades has been the increasingly important role consumption has played as a source of, and way to express, social identity. According to Miller (1995), the impersonality and anonymity of highly rationalized institutions characteristic of modernity place people in a secondary position in relation to the means of production. As people identify less with the institutions that produce goods and services, they come increasingly to identify themselves as consumers who use their consumption of commodities to create specificity and identity. But the so-called 'consumer society' is not simply the outcome of alienation from the institutions and production processes of modernity. Neo-liberal states have instituted a number of strategies to redefine citizenship, and its associated rights and responsibilities, in terms of individualized consumer choices. Moving away from notions of universal rights of access to basic services and participation in collective deliberation and decision-making, neo-liberal states have privatized public services that served to collectivize consumption of goods such as health care and subjected these services to the discipline of 'the market' (Dowding and Dunleavy 1996).

The language of 'consumer choice' and 'deregulation' accompanying neo-liberalism obscures the relations of power implicated in food production-consumption networks. As Abercrombie (1994) points out, the expression of agency by consumers is heavily dependent on the expertise, skills and knowledge they possess in relation to particular commodities. While survey results point consistently towards generally negative attitudes to the new biotechnologies and positive attitudes towards organics (Juanillo 2001; Lockie et al. 2002; Norton 2001), they also point towards high levels of confusion and uncertainty (Lockie et al. 2002; Wansink and Kim 2001). Sources of confusion include the lack of widely understood labelling systems (Sloan 2002; Vladicka and Cunningham 2001); the novel, complex and rapidly changing nature of gene technologies (Norton and Lawrence 1996; Roberts 1994; US FDA 2000), and misleading and exaggerated claims from proponents and opponents of GE regarding the environmental, social and health impacts of both GM and organic food and production systems (Wansink and Kim 2001). Although bewildered by the variety of messages emanating from both public and private agencies or interest groups, food consumers have a tendency to approve less of GM foods as they are provided with more information on them (Norton and Lawrence 1996; Scholderer and Frewer 2003). There is also evidence that the more food consumers are motivated by a concern for the naturalness of foods the more likely they are to shun GM foods and to consume higher levels of organic foods (Lockie et al. in press a, b). In turn, the level of concern which people demonstrate towards the consumption of 'natural' foods is determined primarily by the level of responsibility they take for food purchase and preparation within their households and by their gender.

We are particularly interested in this chapter in relations of power that are based on the deployment of 'technologies of the self' by states and others to influence individual behaviour in relation to food consumption and risk

(Bunton and Petersen 2002). Those technologies of interest include both legislative techniques, such as product labelling regulations, and non-legislative techniques, such as public relations and education campaigns. Importantly, the deployment of these technologies across sectors is uneven, and characterized by considerable compromise and opportunism in the pursuit of particular political goals.

From the perspective of 'consumer citizenship', state support for product labelling may be seen as logically consistent and thereby rational (Jackson 2000). However, the issue of labelling must be seen in relation to other imperatives that continue to confront neo-liberal states such as the need to accommodate the interests of large fractions of capital. Product labelling has thus been a source of considerable conflict both in the GE and organic sectors, albeit for very different reasons. These differences stem not only from the different characteristics of the two industries, but from the fact that while labelling of organic foods may be seen as positive (i.e. emphasizing desirable product characteristics), labelling of GM foods is seen largely as negative (i.e. as somewhat akin to health warnings on tobacco products). The labelling debate in relation to GM foods has been characterized, in general terms, by a biotechnology sector on the one hand which, with the support of most governments, is deeply opposed to labelling those foods that contain GM ingredients and, on the other, a coalition of opposition groups including consumer advocacy, environmental and religious groups for whom labelling is fundamental to consumers' 'right to know' what they are eating. Opponents of labelling have used the concept of 'substantial equivalence', discussed above, to argue that GM labelling would discriminate against such products on the basis of unscientific public hysteria (Goetzl 1999). Segregation and labelling of GE crops would, they argue, add substantial costs and compromise economies of scale (Alden and Landes 2002; Chandrashekhar 2003) while fuelling ignorance and fear over the dangers of GM food (Jones 2002). The USA is representative of this perspective with labelling required only on products that include changes in nutrients or that introduce 'unknown allergens'. Even then, this does not require the product to carry a GM label but to be labelled as to the nutritional change or allergen present (Hansen 2001: 253).

Survey results, however, consistently indicate widespread public support for GM labelling, even in the event that this increases food costs (Chandrashekhar 2003; Hansen 2001; Norton and Lawrence 1996; Scully 2003). The initial response of the GE sector, and the governments that support it, to these concerns was to commence public relations campaigns to educate the public as to the benefits of genetic engineering. In 1999 to 2000, for example, the Australian government formed Biotechnology Australia to provide 'balanced and factual' information on biotechnology to the Australian community. Similarly, in 2001 the US biotech industry launched a US$50 million public relations campaign aimed at preventing the public backlash against GM foods witnessed elsewhere (Hindmarsh and Lawrence

2001). As Hindmarsh and Lawrence (2001: 26) point out, a 'plethora of information packages is now visible on the Internet and in mailouts particularly targeted to younger, potentially more accepting, high-school audiences.' Public relations campaigns have been designed, in many instances, not simply to present gene technologies in a positive light, but to confront those sceptical of these technologies with moral dilemmas regarding the food security and environmental consequences of not pursuing a GM future. Nevertheless, these campaigns have been considerably less successful than their proponents may have hoped. Exaggerated claims regarding the ability of GE to address food security, nutritional and environmental issues have contributed to a widespread perception among food consumers that the benefits of GE are targeted largely at producers and distributors, that something must be wrong with GM, since proponents are trying so hard to convince them otherwise, and that the majority of institutions involved in biotechnology cannot be trusted as sources of reliable information (Juanillo 2001; Wansink and Kim 2001).

Combined with persistent political protest, public support for product labelling has contributed to the introduction of stricter regimes outside the USA, most notably in the European Union where, from April 2004, all foods containing GE ingredients over a threshold level of 0.9 per cent have required labelling (FSA 2003). Importantly, this has not necessarily meant a retraction of state support for the GE sector. While some concessions have been made to public consultation over the regulation of GE and GM food production, this remains largely the domain of expert institutions charged with the scientific determination of risk. Labelling products allowed by these institutions may be interpreted as a means of securing trust in risk-based regulatory regimes while leaving ethical and social considerations to the market. Ultimately, the choice over whether or not to purchase GM foods remains essentially the only way that most people may participate in the process of approving and marketing GM products (Newell 2002).

While there is considerable, and often acrimonious, debate within the organic industry over the specific content of organic certification guidelines, there are few questions regarding the necessity of labelling to reassure consumers that the foods they purchase have been grown and processed using certified organic practices. Nevertheless, despite recent growth in the organic food market, Vladicka and Cunningham (2001) argue that the proliferation of eco-labels and the lack of an internationally accepted definition of 'organic' could contribute to an erosion of consumer confidence, with inconsistency in standards and inspection procedures already issues within the EU. Confusion over the meaning and credibility of labels has left many 'consumers wary of the content and contributions of natural, preservative free, and organic foods' (Sloan 2002: 32). Independent certification organizations have thus lobbied governments to legislate against the use of 'organic' labels on products that do not meet accepted criteria and to help resolve persistent differences between certification bodies (Sligh and Christman 2003). Furthermore, as trade in organic foods expands, moves have been made to harmonize existing standards – which are generally written and regulated by nationally based organizations – at the international level (Sligh and Christman 2003). At the same time, initiatives are underway to foster debate within the organic industry over whether organic production standards should be expanded to include social considerations such as the fair treatment of labour (Henderson *et al.* 2003). In contradistinction to the GE industry, which has attempted to enrol people as consumers of GM foods unconsciously by not labelling their products, the organic industry has sought to offer clearly identifiable choices and thus opportunities for the expression of 'consumer citizenship'.

An important feature of debates over labelling is that they locate questions of power in the relationships between food consumers, producers, state agencies and life science companies. With many consumers complaining that they feel powerless in the face both of a plethora of conflicting discourses regarding the attributes of organic and GM foods, and a GE industry that 'contaminates' foods in ways that cannot be seen, smelt or tasted (Lockie 2002), the value of labelling to the expression of consumer agency and citizenship seems obvious. However, it should not be assumed either that these are the only power relationships implicated in the consumption of GM and organic foods, or that all food consumers wish to express agency in this manner. One of the outcomes of confusion over the genuine attributes of foods has been the opportunity provided to retailers and other actors to exert influence over consumers by assuming the role of mediating conflicting meanings, and the discourses that frame them, on consumers' behalf (Dixon 2002). Leading supermarket chains in the UK have sought to position themselves as responsible corporate citizens by marketing a wide range of organic fruit and vegetables and removing all GM ingredients from their own-brand products (Burch *et al.* 2001). These actions have not been based on any simple concern for consumer well-being but on a capacity – enabled by psychological research and sophisticated knowledge of spending patterns (Dixon 2002) – to engage with consumers in a calculated manner and thus to make connections between the active choices of consumers, and the attributes, anxieties and assurances represented by products (Miller and Rose 1997). That supermarkets pre-empted legislative moves within the EU to institute stricter labelling regimes in the GE and organic sectors is consistent with the pre-emptive approach they have taken in other areas of Quality Assurance (Lockie 1998) as part of a broader strategy to foster consumer trust, and reduce exposure and liability to food-borne risks (Pearce and Hansson 2000). By (1) concentrating ownership within the retail sector, (2) expanding sales of own-brand products, and (3) representing themselves to potential customers and others as champions of consumers' interests, large supermarket chains have captured an increasingly influential role in the mobilization of food networks that exclude GM ingredients.

Conclusion

A number of very different challenges are faced in mobilizing people as consumers of GM and organic foods. Food consumers have demonstrated active resistance to the application of technologies, including genetic engineering, that are seen to compromise the 'naturalness' of food. Despite the efforts of the GE sector to 'educate' the public as to the benefits of the technology, it seems that the more people know about GM food and the more responsibility they take for meeting the food needs of their families, the more resistant they become to the consumption of GM foods. In the case of organics, acceptability of the product is seen as quite high, but price premiums, limited and inconsistent availability, and confusion over the real attributes of organic foods represent significant constraints to the expansion of sales. Yet, with both industries experiencing rapid growth, it is clear that the mobilization of people within food networks involves rather more complicated relationships of power than those implicated in notions of 'consumer demand' and 'consumer citizenship'.

Even though few actors may be in a position to force people to consume particular foods, a range of strategies are used to govern consumption 'at a distance'. At first glance, the most crude of these would appear to have been the exaggerated claims of public relations campaigns and attempts to hide the GM status of foods from their consumers. Nevertheless, the successful establishment of scientific risk assessment – whereby risk must be proven rather than safety – as the dominant mode of ordering for the approval of GM food production and sale has marginalized social and ethical questions within regulatory processes and relegated them to the consideration of 'the market'. Where labelling of GM foods is not required, one of the few viable options for 'consumer citizens' wishing to avoid unwitting ingestion of GM foods has been to purchase only certified organic foods. Thus while the visions of nature and agro-ecology embodied in GE and organic networks could hardly be more different, the expansion of the new biotechnologies has, in turn, fostered expansion of organics. It is certainly tempting, in this context, to interpret the tightening of food labelling regulations in the EU and elsewhere as victories for public protest and consumer sovereignty, but at least two other dynamics within these power relationships must also be considered. First, even where labelling of GM ingredients is required, scientific regulatory agencies have maintained their positions as 'centres of calculation' responsible for speaking on behalf of the genes, animals and plants that must be mobilized within food networks, and risk assessment remains the dominant mode of ordering. Second, supermarket retail chains have sought to establish themselves as 'centres of calculation' responsible for speaking on behalf of consumers and, with a conception of risk based more on the avoidance of liability, have undoubtedly played a significant role in recent changes to regulatory regimes. Ultimately, it may turn out to be the knowledge of consumer spending patterns, desires and so on held by retailers that wields most influence in the governing of consumption.

References

Abercrombie, N. (1994) 'Authority and consumer society', in R. Keat, N. Whiteley and N. Abercrombie (eds) *The Authority of the Consumer*, London: Routledge.

Alden, E. and Landes, D. (2002) 'GM food industry gears up campaign against labels: an initiative on genetically-engineered ingredients could result in significant losses for the biotechnology industry', *Financial Times*, London, October.

Babninard, J. and Josling, T. (2001) 'The stakeholders and the struggle for public opinion, regulatory control and market development', in G. Nelson (ed.) *Genetically Modified Organisms in Agriculture: Economics and Politics*, San Diego, CA: Academic Press.

Black, J. (1998) 'Regulation as facilitation: negotiating the genetic revolution', *Modern Law Review*, 61, 5: 621–660.

Brac de la Perriere, R. and Seuret, F. (2000) *Brave New Seeds: The Threat of GM Crops to Farmers*, London: Zed Books.

Bunton, R. and Petersen, A. (2002) 'Genetics, governance and ethics', *Critical Public Health*, 12, 3: 201–206.

Burch, D., Lyons, K. and Lawrence, G. (2001) 'What do we mean by "green"? Consumers, agriculture and the food industry', in S. Lockie and B. Pritchard (eds) *Consuming Foods, Sustaining Environments*, Brisbane: Australian Academic Press.

Burchell, G. (1993) 'Liberal government and techniques of the self', *Economy and Society*, 22, 3: 267–282.

Callon, M. and Law, J. (1995) 'Agency and the hybrid *Collectif*', *South Atlantic Quarterly*, 94, 2: 481–507.

Campbell, H. (2004) 'Organics ascendent: curious resistance to GM', in R. Hindmarsh and G. Lawrence (eds) *Recoding Nature: Critical Perspectives on Genetic Engineering*, Sydney: University of New South Wales Press.

Chandrashekhar, G. (2003) 'Consumers wary of biotech foods, says US study: an experimental auction has shown that people, on average, are willing to pay 17–21 cents per unit more to purchased plain-labelled food than "GM-labelled" food', *Businessline*, April.

Clapp, J. (2002) 'Transnational corporate interests and global environmental governance: negotiating rules for agricultural biotechnology and chemicals', presented to *International Studies Association Annual Meeting*, New Orleans, 24–27 March.

Dixon, J. (2002) *The Changing Chicken: Chooks, Cooks and Culinary Culture*, Sydney: University of New South Wales Press.

Dowding, K. and Dunleavy, P. (1996) 'Production, disbursement and consumption: the modes and modalities of goods and services', in S. Edgell, K. Hetherington and A. Warde (eds) *Consumption Matters: The Production and Experience of Consumption*, Oxford: Blackwell.

FSA (2003) *Date Set for New GM Food Regulation*, Food Standards Agency, UK, available at http://www.foodstandards.gov.uk/news/newsarchive/new_reg.

Foucault, M. (1991) 'Governmentality', in G. Burchell, C. Gordon and P. Miller (eds) *The Foucault Effect: Studies in Governmentality*, Hemel Hempstead: Harvester-Wheatsheaf.

Goetzl, D. (1999) 'Consumers open to biotech: marketers urged to stress benefits of engineered foods', *Advertising Age*, 70: 30.

Goodman, D., Sorj, B. and Wilkinson, I. (1987) *From Farming to Biotechnology: A Theory of Agro–Industrial Development,* Oxford: Blackwell.

Hansen, M. (2001) 'Genetically engineered food: make sure it's safe and label it', in G. Nelson (ed.) *Genetically Modified Organisms in Agriculture: Economics and Politics*, San Diego, CA: Academic Press.

Henderson, E., Mandelbaum, R., Mendieta, O. and Sligh, M. (2003) *Toward Social Justice and Economic Equity in the Food System: A Call for Social Stewardship Standards in Sustainable and Organic Agriculture*, Pittsboro, NC: Rural Advancement Foundation International – USA.

Higgins, V. (2001) 'Calculating climate: "advanced liberalism" and the governing of risk in Australian drought policy', *Journal of Sociology*, 37, 3: 299–316.

Higgins, V. and Lockie, S. (2001) 'Getting big and getting out: government policy, self-reliance and farm adjustment', in S. Lockie and L. Bourke (eds) *Rurality Bites: The Social and Environmental Transformation of Rural Australia*, Sydney: Pluto Press.

Hindess, B. (1996) *Discourses of Power: From Hobbes to Foucault*, Oxford: Blackwell.

Hindmarsh, R. and Lawrence, G. (2001) 'Bio-utopia: future natural?' in R. Hindmarsh and G. Lawrence (eds) *Altered Genes II: The Future?* Melbourne: Scribe.

Ho, M-W. (1998) *Genetic Engineering: Dream or Nightmare? The Brave New World of Bad Science and Big Business*, Bath: Gateway Books.

Jackson, D. (2000) 'Labeling products of biotechnology: towards communication and consent', *Journal of Agricultural and Environmental Ethics*, 12: 319–330.

Jones, R. (2002) 'Don't feed consumer fears', *American Vegetable Growers*, October: 50–52.

Juanillo, N. (2001) 'The risk and benefits of agricultural biotechnology: can scientific and public talk meet?', *The American Behavioural Scientist*, 44, 8: 1246–1266.

Latour, B. (1987) *Science in Action: How to Follow Scientists and Engineers Through Society*, Cambridge, MA: Harvard University Press.

Latour, B. (1993) *We Have Never Been Modern*, Cambridge, MA: Harvard University Press.

Latour, B. (1999) 'On recalling ANT', in J. Law and J. Hassard (eds) *Actor Network Theory and After,* Oxford: Blackwell.

Law, J. (1991) 'Introduction: monsters, machines and sociotechnical relations', in J. Law (ed.) *A Sociology of Monsters: Essays on Power, Technology and Domination*, London: Routledge.

Law, J. (1994) *Organizing Modernity*, Oxford: Blackwell.

Law, J. (1999) 'After ANT: complexity, naming and topology', in J. Law and J. Hassard (eds) *Actor Network Theory and After,* Oxford: Blackwell.

Lemke, T. (2001) 'The birth of bio-politics: Michel Foucault's lecture at the Collége de France on neo-liberal governmentality', *Economy and Society*, 30, 2: 190–207.

Levidow, L. and Carr, S. (2000) 'Unsound science? Transatlantic regulatory disputes over GM crops', *International Journal of Biotechnology*, 2, 1/2/3: 257–273.

Levidow, L. and Tait, J. (1995) 'The greening of biotechnology: GMOs as environment-friendly products', in I. Moser and V. Shiva (eds) *Biopolitics: A Feminist and Ecological Reader on Biotechnology*, London: Zed Books.

Lockie, S. (1998) 'Environmental and social risks, and the construction of "best-practice" in Australian agriculture', *Agriculture and Human Values*, 15, 3: 243–252.

Lockie, S. (1999) 'The state, rural environments and globalisation: "action at a distance" via the Australian Landcare Program', *Environment and Planning A*, 31, 4: 597–611.

Lockie, S. (2002) '"The invisible mouth": mobilizing 'the consumer' in food production-consumption networks', *Sociologia Ruralis*, 42, 4: 278–294.

Lockie, S., Lawrence, G., Lyons, K. and Grice, J. (in press a) 'Natural foods and biotechnologies: a path analysis of factors underlying support or opposition to biotechnology among Australian food consumers', *Food Policy*.

Lockie, S., Lyons, K., Lawrence, G. and Grice, J. (in press b) 'Choosing organics: a path analysis of factors underlying the selection of organic food among Australian consumers', *Appetite*.

Lockie, S., Lyons, K., Lawrence, G. and Mummery, K. (2002) 'Eating green: motivations behind organic food consumption in Australia', *Sociologia Ruralis*, 42, 1: 20–37.

Miller, D. (1995) 'Consumption as the vanguard of history: a polemic by way of an introduction', in D. Miller (ed.) *Acknowledging Consumption: A Review of New Studies*, London: Routledge.

Miller, P. and Rose, N. (1990) 'Governing economic life', *Economy and Society*, 19, 1: 1–31.

Miller, P. and Rose, N. (1997) 'Mobilizing the consumer: assembling the subject of consumption', *Theory, Culture and Society*, 14, 1: 1–36.

Newell, P. (2001) 'Biotechnology for the poor?', *Science as Culture*, 10, 2: 249–254.

Newell, P. (2002) *Biotechnology and the Politics of Regulation*, IDS Working Paper 146, Brighton: Institute of Development Studies.

Norton, J. (2001) 'Biotechnology to the rescue? Can genetic engineering secure a sustainable future for Australian agriculture?', in S. Lockie and L. Bourke (eds) *Rurality Bites: The Social and Environmental Transformation of Rural Australia*, Sydney: Pluto Press.

Norton, J. and Lawrence, G. (1996) 'Consumer attitudes to genetically-engineered food products: focus group research in Rockhampton, Queensland', in G. Lawrence, K. Lyons and S. Momtaz (eds) *Social Change in Rural Australia*, Rock-hampton, QLD: Rural Social and Economic Research Centre, Central Queensland University.

Pearce, R. and Hansson, M. (2000) 'Retailing and risk society: genetically modified foods', *International Journal of Retail and Distribution Management*, 28, 11: 450–458.

Pew Initiative on Food and Biotechnology (2004) *Feeding the World: A Look at Biotechnology and World Hunger*, Washington, DC: Pew Initiative on Food and Biotechnology.

Roberts, M. (1994) 'A consumer view of biotechnology', *Consumer Policy Review*, 4: 99–104.

Sahota, A. (2004) 'Overview of the global market for organic food and drink', in H. Willer and M. Yussefi (eds) *The World of Organic Agriculture: Statistics and Emerging Trends*, Bonn: International Federation of Organic Agriculture Movements.

Scholderer, J. and Frewer, L. (2003) 'The biotechnology communication paradox: experimental evidence and the need for a new strategy', *Journal of Consumer Policy*, 26, 2: 125–157.

Scully, J. (2003) 'Genetic engineering and perceived levels of risk', *British Food Journal*, 105, 1/2: 59–77.

Sligh, M. and Christman, C. (2003) *Who Owns Organic? The Global Status, Prospects and Challenges of a Changing Organic Market*, Pittsboro, NC: Rural Advancement Foundation International – USA.

Sloan, E. (2002) 'The natural and organic foods marketplace', *Food Technology*, 56, 1: 27–37.

US FDA (2000) *Report on Consumer Focus Groups on Biotechnology*, Washington, DC: United States Food and Drug Administration.

Vladicka, B. and Cunningham, R. (2001) *Snapshot: Organics, A Profile of the Organic Industry and Its Issues*, Alberta: Strategic Information Services Unit, Agriculture, Food and Rural Development.

Wansink, B. and Kim, J. (2001) 'The marketing battle over genetically modified foods: false assumption about consumer behaviour', *The American Behavioural Scientist*, 44, 8: 1405–1417.

Yussefi, M. (2004) 'Development and state of organic agriculture worldwide', in H. Willer and M. Yussefi (eds) *The World of Organic Agriculture: Statistics and Emerging Trends*, Bonn: International Federation of Organic Agriculture Movements.

11 Animals and ambivalence

Governing farm animal welfare in the European food sector

Mara Miele, Jonathan Murdoch and Emma Roe

Introduction

That humans exploit animals, often in cruel ways, is not open to doubt. Neither is there any doubt that responsibility for exploitation and cruelty lies unambiguously on the human side of any human–animal divide. For this reason, relations between humans and animals may be described as profoundly *asymmetrical* (Schiktanz 2004: 2). Asymmetry results whenever animals are confined for human purposes, for instance, in farms, zoos and homes. As Schiktanz (2004: 2) puts it, 'the animal itself has usually no opportunity to force its necessities – everything depends on the good will of the human "owner".' Asymmetric relations are apparently inevitable, especially in the agricultural domain where billions of animals are raised for slaughter in heavily industrialized and mechanized systems of production (Fiddes 1992; Rifkin 1992; Stassart and Whatmore 2003).

Yet asymmetry remains troubling for many humans. Thus as the exploitation of animals for food becomes more intense, so a greater need for regulation seemingly arises. The emergence of animal welfare legislation generates, however, another key dynamic in human–animal relations – *ambivalence*. As Schiktanz (2004: 2) notes, 'the reason for being ambivalent is that on the one hand a specific animal can be individually and compassionately loved and on the other hand various animal species are intensively used in a socio-economic context.' This raises a problem of 'nearness' and 'distance'; that is:

> it reflects the distinct situations of killing animals for food: thus killing companion animals for food reasons is absolutely taboo; whereas for farm animals there are rules depending on the lifecycle of the animal, wild animals are killed in particular seasons and exotic animals wouldn't be used as a food resource at all.
>
> (Schiktanz 2004: 3)

In short, while we feel some kind of connection to animals – meaning they should not be killed or should only be killed in certain ways and at certain times – we also recognize a distance between ourselves and animals – meaning they should be killed so that we can eat.

In this chapter we look a little more closely at 'asymmetry' and 'ambivalence' in the food sector. In particular, we focus on how they influence the construction of animal welfare regulation in Europe and the UK. Animal welfare concern in Europe, as Montanari (1996) indicates, may be traced back to the Victorian period when animal welfare societies began to reflect widespread disquiet over animal treatment (Franklin 1999). These societies ensured that welfare issues remained prominent in Europe throughout the twentieth century – especially as the agricultural industry intensified its animal-based production practices in the post-Second World War era.[1] The recent spate of food scares has brought consumer concerns over farmed animals even more firmly to the fore. Once the conditions of animal production were revealed to the general public (by, for instance, the BSE crisis), anxieties over consumer health were translated into anxieties over animal welfare (Franklin 1999). Thus, in the mid-1990s, around one million people signed a Compassion In World Farming (CIWF) petition for animals to be recognized as 'sentient beings' in European legislation (Watts 1999). It was argued that this new status for animals would bring them enhanced welfare benefits and that these benefits would translate into safer food (Rollin 1995, 2004).

Animal welfare has been creeping up the European political agenda, and it has now given rise to a number of differing regulations and mechanisms of governing. As a result, it forms a key aspect of the agricultural governing system in Europe and elsewhere. In this chapter we take the growing significance of animal welfare as a starting point for considering how farm animals are being regulated in two main arenas. First, we provide an overview of animal welfare legislation in the European Union (EU). We briefly identify the main laws and regulations surrounding welfare and some of the key implications that stem from their adoption. As we shall see, the definition of animal welfare has traditionally been interpreted in the EU as the cluster of external parameters needed to ensure the higher production of farm animals (Spedding 2000). However, during the past twenty years or so this 'productionist' approach has run in parallel with a new concern for the 'global health' of the animal, meaning the total positive psycho-physical conditions that ensure the survival of sentient life (Broom 1991; Wilkins 1997). We shall argue that these two conceptions of the animal remain current in EU legislation, thus bestowing on the farmed animal a profoundly ambivalent status.

Second, we consider how welfare laws and regulations are administered in the arena of the nation state. We suggest that the UK constitutes an instructive case study: concern with animal welfare is a long-standing issue in this country (Harper and Henson 1998; Harrison 1964); moreover, the UK has suffered from acute animal disease problems, notably Foot and Mouth disease (which cost the British taxpayer around £15 billion in 2001 to 2002), BSE (which has so far killed around 100 people), classical swine fever (which has become a recurring concern in the UK pig industry), and food poisoning epidemics (such as salmonella and campylobactor). The prevalence of these diseases has made the UK government especially sensitive to animal health and welfare issues, and we consider how it has sought to implement a range of welfare measures. In so doing, we trace the networks of actors that facilitate welfare regulation in the nation state context.

Before examining the governing of welfare in the European policy arena it should be noted that the analysis of animal welfare which follows is informed implicitly by a Foucaultian perspective. In particular, it considers whether animal welfare now constitutes a new regime of 'governmentality'. Foucault (1991) uses this term to refer to the collective ways of thinking that underpin particular governmental strategies. In his view all modes of regulation depend on modes of 'representation'; that is, specific ways of depicting the domain to be governed. In general terms, modes of representation make given domains amenable to political deliberation. They also tend to define common vocabularies that permit the mobilization of diverse social and political actors. The adoption of shared vocabularies enables associations to be formed between a variety of agents dispersed in space and time. As Miller and Rose (1990: 6) put it:

> departments of State, pressure groups, academics, managers, teachers, employees, parents – whilst each remains, to a greater or lesser extent, constitutionally distinct and formally independent ... can be enrolled in a governmental network to the extent that it can translate the objectives and values of others into its own terms, to the extent that the arguments of another become consonant with and provide norms for its own ambitions.

In what follows we consider, first, whether the welfare of farm animals has now become a discrete and defined 'object' of governing; that is, we investigate how animal welfare has been delimited as a governmental problem. As we shall see, problems have arisen in the governmental arena in defining 'the animal' (i.e. is the animal a 'production machine' or a 'sentient being'?). We argue that these problems of definition reflect a profound ambivalence towards animals, one that generates inconsistencies in prevailing governmentalities. Second, we describe the network of actors that have been mobilized as processes of animal welfare regulation have come into being. As we shall see, the governmental network now bearing upon the welfare problematic is becoming increasingly complex in character: it consists not only of government agencies but also of non-governmental actors and private sector organizations. This regulatory network is essentially working to 'frame' the actions of all those engaged in the food sector using standards, prescriptions and norms of animal welfare practice. In our view, such welfare 'framings' not only constitute an emerging form of governmentality but also comprise an increasingly important part of the system of agricultural governing.

Animal welfare legislation in the EU

A growing number of EU recommendations, laws and treaties aim to regulate the relationship between humans and animals. While European Union law takes a variety of forms – including directives, regulations and decisions – all must ultimately be incorporated into an EU Treaty if they are to become legally binding on member states. Yet, despite the fact that animal welfare is clearly an issue of great concern to many EU citizens (Bennett 1996), European animal welfare associations strongly argue that this concern is not sufficiently reflected in the existing EU Treaties (see www.eurogroup-foranimalwelfare.net). In fact, as we shall see below, there exists a profound ambiguity in EU legislation on animal welfare. On the one hand, the existing legislation sees animals as mere production resources (e.g. agricultural products or animals employed in medical research); on the other hand, it sees animals as entities with a special status and specific legal requirements (e.g. companion animals). This ambiguity underpins many current disputes over farm animals in the EU context.

In the first instance, EU directives dealing with farm animal welfare have been generated principally by the need to establish common rules that will ensure the proper functioning of the internal European market. As Moynagh (2003: 108) points out:

> It is often forgotten that the European Union is a trading body. Though it has grown in breadth and depth, one of its primary roles remains to assure the single market and to ensure free trade in goods and in services. One of the first groups of commodities traded was agricultural goods – of which animals and animal products are an important part. For this reason, veterinary legislation developed earlier than other areas of EU legislation and is generally more comprehensive than legislation dealing with other commodities and substances. There has thus been a considerable degree of harmonization of legislation between Member States in order to ensure that no Member State obtained an unfair advantage. Such harmonization has also covered welfare standards and, in particular, the setting of minimum welfare standards which apply across the EU.

However, it is not only anxieties over competitiveness that have led to the introduction of welfare legislation: European animal welfare associations have lobbied to ensure enhanced welfare is made a basic principle of EU governing. In the negotiation between the animal welfare associations and the EU a key element for discussion has been the scientific knowledge in the field of animal science. Of key importance here has been the Scientific Committee on Animal Health and Welfare (SCAHW) which operated until very recently as a scientific advisory committee of the EU (it has now been replaced by scientific panels under the European Food Standards Agency).

The SCAHW has traditionally been composed of leading scientists in the field of animal health and welfare.

We can identify two main welfare approaches in the advice that the SCAHW has supplied to the European Commission. While these approaches are not mutually exclusive, they do define the welfare of animals in sharply differing ways. The first may be termed the 'environmental approach', since it interprets the welfare of animals as the cluster of external parameters needed to ensure high levels of production. It focuses on the combination of maximum production and minimum cost through the creation of an environment in which animals are easily transformed into food products (see e.g. Kleiber 1961; Mount 1968). The second approach looks at welfare from the perspective of the animal rather than the environment. One strand of animal-centred work examines the ability of animals to adapt to (or cope with) the farm environment (see e.g. Broom 1991, 1996) while another strand seeks to understand how the animal feels about the farm (confinement) situation (see e.g. Dawkins 1980; Duncan and Petherick 1989; Fraser and Duncan 1998). These new scientific findings – especially those that focus upon the animal's likely ability for self-awareness and suffering, and its capacity to feel complex emotions associated with fear, pain and behavioural needs (Blockhuis *et al.* 2003) – have made a profound impact on perceptions of human duties towards animals in terms of limitations to the suffering, deprivation and various distresses connected with animal farming and other forms of animal exploitation. More specifically, they have supported the view that farm animals are 'sentient beings'. Thus, in its advice to the EU Commission – for instance, on slaughtering methods or on transportation issues – the SCAHW has increasingly tended to emphasize animal-centred welfare approaches (Moynagh 2003).

In this way, new conceptions of 'welfare' as an object of governing have emerged in EU circles and these have begun to influence animal welfare legislation, including a Declaration on Animal Welfare in the 1991 Maastricht Treaty and a Protocol on Animal Welfare in the 1997 Amsterdam Treaty (see Figure 11.1). This latter Protocol was an especially important milestone, since it indicated that animal-centred definitions were moving to the fore. The Protocol reads as follows:

> The High Contracting Parties, desiring to ensure improved protection and respect for the welfare of animals as *sentient beings*, have agreed upon the following provision, which shall be annexed to the Treaty establishing the European Community, in formulating and implementing the Community's agricultural, transport, internal market and research policies, the Community and the Member States shall pay full regard to the welfare requirements of animals, while respecting the legislative or administrative provisions and customs of the Member States relating in particular to religious rites, cultural traditions and regional heritage.

The Protocol creates clear legal obligations on EU member states to pay full regard to the welfare requirements of animals and, for the first time, refers to them as 'sentient beings', thereby bestowing special obligations on all those who rear animals. However, while the Protocol seemingly introduces a new rationale for animal welfare regulation, in Annex I (Article 32) of the Treaty animals are still referred to as 'agricultural products'. Thus ambiguity resurfaces. In fact, taken as a whole, the Amsterdam Treaty appears to see animal welfare as a subject that should be encompassed within other EU policy areas, such as the Common Agricultural Policy and the internal market.

The most recent legislation bearing upon animal welfare is the new European Constitutional Treaty, which was agreed on 18 June 2004.[2] Importantly, the Treaty transforms the animal welfare Protocol into a Treaty Article. The Article is to be found in Part III of the Treaty, which is entitled 'The Policies and Functioning of the Union'. The new Article is similar in its wording to the 1997 Protocol and reads:

> In formulating and implementing the Union's agriculture, fisheries, transport, internal market, research and technological development and space policies, the Union and the Member States shall pay full regard to the welfare requirements of animals, as sentient beings, while respecting the legislative or administrative provisions and customs of Member States relating in particular to religious rites, cultural traditions and regional heritage.

The Article has two key elements: first, it reaffirms that animals are 'sentient beings' (this means they cannot be regarded simply as goods or products), and second, it requires the EU and member states, when formulating and implementing EU policies, to pay full regard to the welfare requirements of animals. The new Article relates both to policies that directly affect animals (for example, a proposed directive on cattle welfare) and to policies that may have an indirect impact on animals (such as a new policy on the safety of certain products, which could lead to more animal testing). In the latter case, the Commission is being urged to recognize that it must, as a consequence of the new Article, carry out an 'animal welfare impact assessment' before adopting any new policy. This may be seen as a 'mainstreaming' of animal welfare into general EU policy.

This brief overview indicates that 'animal welfare' is going through a process of refinement as an object of government in the European context. In the early rounds of the governmentalization process, welfare was simply seen as an intrinsic part of the agricultural production system – put crudely, if an animal could grow in line with the production expectations then its welfare was not seriously in doubt. However, as more sophisticated scientific understandings of the plight of animals in modern production systems come to be bolstered by growing societal concerns (articulated by animal welfare

General
Protocol (No 33) to the Treaty establishing the European Community on the protection and welfare of animals (1997, adopted 1 May 1999)

Keeping of animals
Council Directive 98/58/EC of 20 July 1998 concerning the protection of animals kept for farming purposes
Council Directive 88/166/EEC of 7 March 1988 laying down minimum standards for the protection of laying hens kept in battery cages
Council Directive 1999/74/EC of 19 July 1999 laying down minimum standards for the protection of laying hens
Council Directive 91/629/EEC of 19 November 1991 laying down minimum standards for the protection of calves
Council Directive 97/2/EC of 2 January 1997 laying down minimum standards for the protection of calves
Commission Decision 97/182 of 24 February 1997 amending the Annex to Directive 91/629/EEC laying down minimum standards for the protection of calves
Council Directive 91/630/EEC of 19 November 1991 laying down minimum standards for the protection of pigs
Council Directive 2001/88/EC of 23 October 2001 amending Directive 91/630/EEC laying down minimum standards for the protection of pigs
Commission Directive 2001/93/EC of 9 November 2001 amending Directive 91/630/EEC laying down minimum standards for the protection of pigs

Transport of animals
Council Directive 91/628/EEC of 19 November 1991 on the protection of animals during transport and amending Directives 90/425/EEC and 91/496/EEC
Council Directive 95/29 of 29 June 1995 on the protection of animals during transport
Council Regulation (EC) No 1255/97 of 25 June 1997 concerning community criteria for staging points and amending the route plan referred to in the Annex to Directive 91/628/EEC
Council Regulation (EC) 411/98 of 16 February 1998 on additional animal protection standards applicable to road vehicles used for the carriage of livestock on journeys exceeding eight hours
Commission Decision 2001/298/EC of 30 March 2001 amending the Annexes to CONV 842/03 12
Council Directives 64/432/EEC, 90/426/EEC, 91/68/EEC and 92/65/EEC and to Commission Decision 94/273/EC as regards the protection of animals during transport

Slaughter and killing of animals
Council Directive 93/119/EEC of 22 December 1993 on the protection of animals at the time of slaughter and killing

Figure 11.1 Overview of EU regulation for welfare of farm animals

organizations), so more nuanced regulatory initiatives come into being. These more nuanced initiatives take the animal's feelings and emotions into account as well as broad aspects of physiology, ethology and health. In short, they see animals not as production 'machines' (the industry view) but as 'sentient beings' (the scientific view). This new perspective has now been incorporated into EU legislation, beginning with the 1997 Amsterdam Treaty and culminating in the new EU Constitution.

However, before celebrating this shift to an animal-centred approach, we should note that a profound ambiguity over the status of farm animals remains and that this inhibits attempts to stabilize animal welfare as a governmental problem. On the one hand, concerns over competitiveness continue to decree that animals are seen mainly as inputs into ever more efficient agricultural production systems. On the other hand, animals are conceptualized as very distinct entities within such production systems, entities in need of special protection tailored to their status as living, conscious beings. The co-existence of these two views means that EU legislation can be interpreted (by, for instance, member states) in distinct ways; that is to say, it legitimizes the continued exploitation of animals in line with competitiveness concerns, or it upholds the need for the high standards of welfare that are associated with new understandings of animal health and well-being.

National systems of welfare governing: a UK case study

EU legislation needs to be interpreted by member states and it is here that we should expect to find the differing conceptions of welfare identified above coming more fully into view. Indeed, there is clear variation in the application of welfare standards around Europe with Scandinavian countries generally upholding high standards and southern and eastern countries being less concerned with welfare issues.[3] We have chosen to focus in this section on a single country – the UK – that has long displayed high levels of animal welfare concern and which has also been subject to acute food scares associated with intensive systems of animal production. This combination ensures that the UK has come to place some considerable emphasis on animal welfare schemes in recent years (in part, because the agricultural industry is concerned to maintain access to international markets for its animal-based products). The implementation of animal welfare measures in the context of food scares and animal disease problems makes the UK a useful case for study, since we can rather easily identify the key policy networks that now surround welfare policy. A brief investigation of these networks should show whether the ambiguity so evident at the EU level is replicated at the national level.

We must first recognize that in the UK a comprehensive framework of legislation has existed for some time to protect farm animals. The Protection of Animals Act 1911 makes it an offence to cause unnecessary suffering to any domestic or captive animal, while the Agriculture (Miscellaneous Provisions) Act 1968 authorizes agriculture ministers to issue regulations specifying detailed conditions under which livestock must be kept. The UK is also required to implement into domestic law any EU Directives bearing upon this issue. There are currently two EC Directives laying down minimum standards for the welfare of specific farm animals: 91/630/EEC (pigs) and 97/182/EC (calves). These Directives are implemented in the UK by the Welfare of Livestock (Amendment) Regulations 1998. In addition, EU Directive 98/58/EC, which sets minimum standards for the welfare of all farm animals, is implemented through the Welfare of Farmed Animal Regulations, which came into force in 2000. Specific rules on the welfare of laying hens are set down in EU Directive 99/74/EC, and these have been implemented in England through the Welfare of Farmed Animals (England) (Amendment) Regulations 2002 (the Directive and domestic implementing regulations prohibit the use of the barren cages with effect from 1 January 2012).

The UK government not only administers these legal functions but also encourages farmers to adopt high standards of animal husbandry through the publication of specific welfare codes. Although these codes are not directly applicable in law, failure to observe their provisions may be used in support of a prosecution for offences under the 1968 Act. As Barclay and Hughes (1998: 7) put it, 'it is not an offence to infringe the terms of the codes of practice, but failure to conform to them can be cited in court as evidence of cruelty in the case of a prosecution for cruelty to animals.' In the main, the codes are enforced by the State Veterinary Service (SVS), which visits farm premises to check the welfare of livestock, and investigates complaints and allegations that welfare requirements have been infringed. Through this close monitoring of on-farm welfare practice the SVS plays a vital role in bringing a welfare governmentality into being at the local level. Independent advice to government in the field of animal welfare standards is provided by the Farm Animal Welfare Council (FAWC), a standing committee established in 1979. Its terms of reference are to keep under review the welfare of farm animals and to advise the government of any legislative or other changes that may be necessary. The council has freedom to investigate any topic falling within its remit and to publish its advice independently (see http://www.fawc.org.uk/). The majority of FAWC recommendations are implemented by legislation and welfare codes.

The UK government not only oversees the policing of the agricultural industry but also mobilizes welfare discourses in order to encourage farmers to monitor their own conduct in welfarist terms. It issues advisory booklets on specific welfare issues (e.g. lameness, heat stress, condition scoring, lamb/calf survival, poultry welfare), and also runs advisory meetings and workshops through its agricultural extension services. Through these discursive mobilizations, the UK government hopes to spread a welfarist ethos through the agricultural industry. This ethos is also evident in the recently published *Animal Health and Welfare Strategy for Great Britain* (DEFRA 2004). The new governmental initiative has a number of broad aims, including:

- That animals kept for food, farming, sport, companionship, entertainment and in zoos should be treated humanely.
- That the disease status of animals in the UK should remain among the highest in the world so as to allow trade in animals and animal products.
- That the costs of animal welfare measures should be appropriately balanced between industry and taxpayer or consumer.
- All disease emergencies should be dealt with effectively and swiftly.
- Consumers should come to value the confidence they have in food produced to high welfare standards.

Arguably, the overriding aim of this strategy is to ensure (following the recent outbreaks of Foot and Mouth disease and BSE) that national and international markets remain open to British animal products. As the strategy document puts it, 'consumers have fundamental expectations about acceptable levels of animal health, the safety of the food they eat, and that standards of animal welfare appropriate to a modern society have been met' (DEFRA 2004: 28).

Another striking feature of the new approach is the emphasis it places on 'partnership' between various industry 'stakeholders'. As the document puts it:

> This strategy does not provide a magic wand to solve all the problems affecting the health and welfare of our animals. But it sets a framework and direction for a partnership between all of us who have the capacity or the responsibility to influence the health and welfare of animals. Such a partnership is crucial if we are to ensure that the continually evolving threats to animal health and welfare are effectively identified, assessed and acted upon. We hope that all who read this strategy will rise to the challenge with enthusiasm, dedication and shared commitment.
>
> (DEFRA 2004: 12)

In identifying appropriate partners the strategy document refers to 'third sector' organizations that run farm welfare assurance schemes. A leading exponent of this 'third way' approach to welfare regulation is the Royal Society for the Prevention of Cruelty to Animals (RSPCA). Since 1994, the RSPCA has run the Freedom Food scheme. It now includes around 150 million animals housed in approximately 1,500 production units. This scheme effectively implements the RSPCA species-specific welfare standards on farms and among hauliers and abattoirs. In general terms, the standards are based on the 'five freedoms' defined by the government's animal welfare advisory body, the Farm Animal Welfare Council (see Figure 11.2). Before a farmer, haulier or abattoir can join the Freedom Food scheme, an RSPCA-approved assessor must carry out a detailed audit on the farm or on the business premises to ensure that these 'freedoms' are encompassed within the production or transportation system. Once enrolled in the scheme, members are subject to regular reassessments to ensure that the 'freedoms' are being promoted on the farm or in the livestock business. In addition, the RSPCA's farm livestock officers carry out random spot checks to help ensure that the standards are being adhered to (see Figure 11.2).

Underpinning the five freedoms are explicit criteria tailored to each species and each production system. For instance, the RSPCA produces guidelines for laying hens which stipulate that 'hens must have access to nutritious food at all times each day, except when required by the attending veterinary surgeon', with 'particular attention ... given to the provision of food and water in areas frequented by subordinate hens'. Producers 'must have a written record of the nutrient content of the feed, as declared by the feed compounder, and must make it available to the Freedom Food assessor and RSPCA farm livestock officer.' When it comes to the environment, it is stipulated that 'all hens must have sufficient freedom of movement to be

---

The welfare of an animal includes its physical and mental state and we consider that good animal welfare implies both fitness and a sense of well-being. Any animal kept by man must, at least, be protected from unnecessary suffering.

The five freedoms

We believe that an animal's welfare, whether on farm, in transit, at market or at a place of slaughter should be considered in terms of **'five freedoms'**. These freedoms define ideal states rather than standards for acceptable welfare. They form a logical and comprehensive framework for analysis of welfare within any system together with the steps and compromises necessary to safeguard and improve welfare within the proper constraints of an effective livestock industry.

1 **FREEDOM FROM HUNGER AND THIRST** – by ready access to fresh water and a diet to maintain full health and vigour.

2 **FREEDOM FROM DISCOMFORT** – by providing an appropriate environment including shelter and a comfortable resting area.

3 **FREEDOM FROM PAIN, INJURY OR DISEASE** – by prevention or rapid diagnosis and treatment.

4 **FREEDOM TO EXPRESS NORMAL BEHAVIOUR** – by providing sufficient space, proper facilities and company of the animal's own kind.

5 **FREEDOM FROM FEAR AND DISTRESS** – by ensuring conditions and treatment which avoid mental suffering.

Source: UK Farm Animal Welfare Council (see: www.fawc.org.uk).

Figure 11.2 The 'five freedoms'

able, without difficulty, to stand normally, turn around and stretch their wings' and 'all hens must have sufficient space to be able to perch or sit quietly without repeated disturbance'. On health, producers must put in place a written Veterinary Health Plan with a veterinary surgeon (see RSPCA 2003: 2–3). They must also keep detailed health records, including details of any medication. Each of the five freedoms is fleshed out in this fashion for each species.

In short, the Freedom Food scheme requires producers to apply a set of tight regulations on the treatment and maintenance of farm animals. The basic aim of the scheme is to provide an assurance to the consumer that animal welfare standards have been met at all stages in the supply chain. As the RSPCA website puts it:

> Consumers can be confident that before products can appear on the supermarket shelves bearing the Freedom Food trademark, traceability must be established through the supply chain. If the farmer is a chicken producer, for example, the hatchery from which they were sourced must be accredited. The haulier who delivered them to the farm and who will eventually take them on to the abattoir must have been successfully assessed, and the abattoir itself must also satisfy all the RSPCA welfare conditions.
>
> (see www.rspca.org)

This brings us to another obvious partner in the governmental pursuit of higher welfare standards – the retail sector. As the government's new strategy document puts it, 'retailers and their customers can specifically support and reward farmers who invest in standards of animal health and welfare that exceeds the acceptable norm' (DEFRA 2004: 28). Likewise, Young (2004: 64) notes that the huge buying power of supermarkets 'means they can move quickly and decisively – perhaps more so than political decision makers – on food standard issues including animal welfare'.

Despite some considerable variation in the attitudes of the major UK supermarkets to animal welfare issues, there is some evidence that at least a minority of retailers are taking the issue seriously. For instance, in 1997 Marks & Spencer became the first major UK retailer to sell exclusively free-range eggs. Then in September 2002 it became the first retailer to use only free-range eggs in all food products (according to the company this covers 250 million eggs a year, laid by 700,000 chickens – see www.marksand-spencer.co.uk). In addition, the Marks & Spencer Select Farm scheme aims to raise welfare standards by ensuring that 'animals will be bred outdoors and benefit from more space and straw bedding, allowing them to live and behave more naturally' (ibid.). Following these moves into welfare-friendly sourcing, Marks & Spencer was awarded the title of 'Compassionate Super-market of the Year' by the campaign group CIWF in 2002. In 2004, however, Waitrose won this title (see CIWF 2002, 2004). It too runs a strict farm assurance scheme which not only ensures that the farm environment is controlled to high standards 'but also provides an audit trail that gives Waitrose the assurance of quality we require' (see www.waitrose.co.uk).

Waitrose and Marks & Spencer are undoubtedly the leading retailers in welfare-friendly food products. There are not such clear commercial market agendas for the sale of welfare-friendly food products among the other UK supermarkets (Tesco, Sainsbury, Asda, Morrisons, Somerfield). As a result they are some way behind the market leaders. Nevertheless, there is some interest in animal welfare. For example, the UK's biggest retailer, Tesco, was an early supporter of Freedom Food, and it is involved in sponsoring various research projects on animal welfare issues including the Food and Animal Initiative (FAI) in Oxford. One project from the FAI has aimed to identify new ways of improving the taste of Tesco Finest's pork products. As a consequence, 'much higher fibre content has been introduced to the pigs' diets. This is beneficial to the intestinal health of the animals and consequently their overall well-being' (www.tesco.com).

It seems, then, that a small number of UK supermarkets are monitoring the animal welfare practices of their various suppliers. In this regard, the supermarkets are also key agents of welfare governmentality (they practice what Marsden *et al.* (2000) describe as 'private interest governance'). However, it should be noted that the supermarkets themselves are being monitored in turn by non-governmental welfare organizations (see Freidberg 2004). One recent initiative of this type was 'The Race To The Top', which was established in 2000 by the Institute for International Environment and Development in order 'to help the major UK supermarkets enhance their social, environmental and ethical policies and performances, through a process of engagement with a variety of civil society organisations' (Fox and Vorley 2004: 20). The assessment process was based on benchmarks of supermarket performance and included an animal welfare component (Lymbery 2000). While the initiative was initially successful in drawing attention to the variable standards of supermarkets, it was unfortunately short-lived and ended in January 2004. As one member of the advisory group noted: 'the consumer and the citizen are generally not the same person, and supermarket companies listen to the former first and the latter a long way second' (quoted in Fox and Vorley 2004: 23). Another group that monitors the supermarkets is CIWF. In 2001 and 2003 the organization produced reports under the title 'Raising the Standard' which assessed the performance of supermarkets only on animal welfare criteria. On the basis of the assessments, the supermarkets were ranked in terms of their animal 'friendliness', with Marks & Spencer and Waitrose gaining the highest scores in the two reports produced so far (see CIWF 2002, 2004).

These various cross-cutting initiatives indicate that the welfare and health of livestock have become issues of increasing public concern in the UK. In response, the government has introduced higher standards of legislation that aim to improve directly the lives of farmed animals. However, the implementation of these standards requires the support of non-governmental 'partners' including farmers, retailers and consumers. In short, the regulation of farm animal welfare is conducted by a complex network of actors, including government agencies, campaign groups and private sector organizations. The linkages between all these actors are often close: the government sees the RSPCA and the supermarkets as key agents in the delivery of its own animal welfare strategy; the RSPCA works through the supermarkets and other retail outlets to ensure that its Freedom Food products reach large numbers of consumers; and supermarkets draw upon the legitimizing powers of groups such as CIWF in order to build up consumer confidence in their own assurance schemes. This integrated network – which Freidberg (2004) calls an 'ethical complex' – is slowly putting in place a new set of animal welfare standards and practices.

Yet, while great efforts are clearly being made to raise animal welfare standards in the UK – prompted mainly by the catastrophic consequences that have followed from the outbreaks of BSE and Foot and Mouth – there are still some unresolved ambiguities within the emerging systems of regulation. For instance, the UK government's own animal welfare strategy seems to be aimed mainly at regaining consumers' confidence and a share of the export market. The RSPCA's Freedom Food scheme adopts a more animal-centred approach. It specifies clear standards and guidelines derived from the needs of the animal itself. The supermarkets appear to occupy an intermediate position: they adopt higher welfare standards in order both to reassure consumers and specify particular market niches for their products (i.e. not all UK supermarkets are competing on higher welfare standards – most are concerned mainly with low prices); however, these higher standards do seem to be having a clear impact at the farm level (Marks & Spencer move into free-range eggs is a shift of some considerable significance given the numbers involved). These differing emphases again indicate that animal welfare as an object of governing is still in the process of clarification: it is still not clear exactly what 'welfare' means in the various regulatory networks that are responsible for its implementation. It is perhaps for this reason that welfare problems continue to bedevil UK agriculture. As DEFRA (2004: 12) admits: 'in 2003 there were 1,610 confirmed bovine TB incidents compared with 720 in 1998'; 'in 1999–2000, a survey of pigs before slaughter showed about 23 per cent were infected with salmonella'; 'in 2003, out of 4,964 farm inspections carried out by the State Veterinary Service, 1,431 (28 per cent) failed to comply with statutory welfare legislation'. In other words, there is still some way to go before the animal welfare network becomes a more effective regime of governmentality.

Conclusion

The preceding pages have shown that animal welfare regulation is a key aspect of agricultural and food regulation. At the EU level a body of legislation is slowly beginning to be assembled so that welfare issues are moving further towards the centre of policy. At present there is a concerted effort by animal welfare organizations and certain national governments to ensure that the definition of animals as 'sentient beings' becomes part and parcel of EU law. Once enshrined in law, it is hoped that new policies and practices towards animals will become more widespread with the effect that animal health and welfare will be markedly improved across EU member states. In this regard, the EU is evidently aiming to turn itself into a zone of enhanced welfare standards (a place where farm animals are routinely seen as 'sentient beings'). In the UK context a similar approach is currently being tried. EU directives are being implemented, and various codes and strategies are emerging that aim to make the UK a welfare-friendly environment. While some of the existing legislation is long-standing, recent legislative initiatives have been put in place in an attempt to overcome the devastating consequences of the BSE and Foot and Mouth outbreaks. It may therefore be assumed that the various EU and national initiatives now coming into force will move the UK towards a greater concern for animal sentience rather than just animal productivity. In other words, a new governmentality of animal welfare seems to be emerging with its own rationalities and technologies of implementation.

Yet, we might question whether the new governmental interest in animal welfare will gel into a coherent welfare regime. While a large number of initiatives are being developed at all scales of government, these are often designed with sharply differing objectives, reflecting perhaps continued ambivalence around 'welfare' as an object of governing. For instance, in the UK case we have seen that welfare measures are introduced for a variety of (not always compatible) reasons including: to keep open international markets for national animal products, to disseminate more animal-friendly methods of production, and to demarcate discrete market niches for retailers. These varied objectives mean that no common means of implementing 'animal welfare' is likely to be adopted in the near future (all the schemes mentioned have rather differing standards and regulations attached). Thus producers and other supply chain actors seem set to remain encompassed within cross-cutting networks, all carrying slightly differing prescriptions, standards and directions. It may be, then, that those who would prefer to slip into the spaces between the networks in order to evade any full engagement with the governmentality of welfare will find plenty of opportunity to do so. Thus, to summarize the situation, and the argument of this chapter, a damaging asymmetry – that is, the incorporation of animals into production systems which cause unnecessary suffering – will be perpetuated by a disabling ambivalence – that is, an unwillingness to recognize the full extent of animal needs and wants. The consequence will be continuing problems of animal health, animal welfare and food quality.

Acknowledgements

This chapter is based upon work carried out in the research project *Welfare Quality*. The project is co-financed by the European Commission within the 6th Framework Programme (contract No. FOOD-CT-2004-506508). The text represents the authors' views and does not necessarily represent a position of the Commission, which will not be liable for the use made of such information. We are grateful to our participants on the project for stimulating many of the ideas discussed above, especially Harry Blockhuis, Bryan Jones, Rony Geers and Isabelle Vessier. However, it must be emphasized that the views expressed here are those of the authors alone.

Notes

1 Exemplified elegantly by the publication in 1964 of Ruth Harrison's seminal book *Animal Machines*.
2 It is due to come into force on 1 November 2006, provided it has been ratified by all the member states.
3 Although this geography of animal welfare may be subject to change as consumers in the South and East become as concerned as consumers in the North of Europe about standards of food quality.

References

Barclay, C. and Hughes, P. (1998) 'Animal welfare', *House of Commons Research Papers*, 98, 12: 1–36.
Bennett, R.M. (1996) 'People's willingness to pay for farm animal welfare', *Animal Welfare*, 5: 3–11.
Blokhuis, H.J., Jones, R.B., Geers, R., Miele, M. and Veissier, I. (2003) 'Measuring and monitoring animal welfare: transparency in the food product quality chain', *Animal Welfare*, 12: 445–455.
Broom, D.M. (1991) 'Animal welfare: concepts and measurement', *Journal of Animal Science*, 69: 4167-4175.
Broom, D.M. (1996) 'Animal welfare defined in terms of attempts to cope with the environment', *Acta Agricultural Scandinavian Supplement*, 27: 22–28.
Compassion in World Farming (CIWF) (2002) *Supermarkets and Farm Animal Welfare: 'Raising the Standard'*, Hampshire: CIWF Trust.
Compassion in World Farming Trust (CIWF) (2004) *Supermarkets and Farm Animal Welfare: 'Raising the Standard'*, Hampshire: CIWF Trust.
Dawkins, M. (1980) *Animal Suffering: The Science of Animal Welfare*, London: Chapman and Hall.
DEFRA (2004) *Animal Health and Welfare Strategy for Great Britain*, London: DEFRA.
Duncan, I. and Petherrick, J. (1989) 'Cognition: the implications for animal welfare', *Applied Animal Behaviour Science*, 24: 24–81.
Fiddes, N. (1992) *Meat: A Natural Symbol*, London: Routledge.
Foucault, M. (1991) 'Governmentality', in G. Burchell, C. Gordon and P. Miller (eds) *The Foucault Effect: Studies in Contemporary Rationalities of Government*, London: Harvester Wheatsheaf.

Fox, T. and Vorley, B. (2004) *Top to Bottom,* London: The Environment Council.

Franklin, A. (1999) *Animals and Modern Culture,* London: Sage.

Fraser, D. and Duncan, I. (1998) 'Pleasures, pains and animal welfare: toward a natural history of affect', *Animal Welfare,* 7: 383–396.

Freidberg, S. (2004) 'The ethical complex of corporate food power', *Environment and Planning D: Society and Space,* 22: 513–531.

Harper, G. and Henson, S. (1998) *Consumer Concerns About Animal Welfare and the Impact on Food Choice EU FAIR-CT98-3678,* Reading: University of Reading.

Harrison, R. (1964) *Animal Machines,* London: Vincent Stuart.

Kleiber, M. (1961) *The Fire of Life – An Introduction to Animal Energetics,* New York: Wiley.

Lymbery, P. (2000) *Module Briefing Paper Animal Welfare,* London: Race to the Top.

Marsden, T., Flynn, A. and Harrison, M. (2000) *Consuming Interests: The Social Provision of Foods,* London: UCL Press.

Miele, M. and Parisi, V. (2001) 'L'Etica del Mangiare, i valori e le preoccupazioni dei consumatori per il benessere animale negli allevamenti: Un'applicazione dell'analisi Means-end Chain', *Rivista di Economia Agraria,* Anno LVI, 1: 81–103.

Miller and Rose (1990) 'Governing economic life', *Economy and Society,* 19, 1: 1–31.

Montanari, M. (1996) *The Culture of Food,* Oxford: Blackwell.

Moynagh, J. (2000) 'EU regulation and consumer demand for animal welfare', *AgBioForum,* 3, 2 and 3: 107–114.

Mount, L.E. (1968) *The Climatic Physiology of the Pig,* London: Arnold.

Rifkin, J. (1992) *Beyond Beef: The Rise and Fall of Cattle Culture,* New York: Dutton Books.

Rollin, B. (1995) *Farm Animal Welfare: Social, Bioethical and Research Issues,* Ames: Iowa State University Press.

Rollin, B. (2004) 'The ethical imperative to control pain and suffering in farm animals', Paper presented to Eursafe Conference, 'Science, Ethics and Society', Leuven, September.

RSPCA (2003) *Welfare Standards for Laying Hens and Pullets,* London: RSPCA.

Schiktanz, S. (2004) 'Ethical consideration of the human–animal relationship under conditions of asymmetry and ambivalence', Paper presented to Eursafe Conference, 'Science, Ethics and Society', Leuven, September.

Spedding, C. (2000) *Animal Welfare,* London: Earthscan.

Stassart, P. and Whatmore, S. (2003) 'Metabolizing risk: food scares and the un/remaking of Belgian beef', *Environment and Planning A,* 35, 3: 449–462.

Watts, M. F. (1999) 'An agenda for reform: Farm Animal Welfare in the European Union', in G. Tansey and J. D'Silva (eds) *The Meat Business: Devouring a Hungry Planet,* London: Earthscan.

Wilkins, D. (1997) *Animal Welfare in Europe,* Wallingford: CAB International.

Young, W. (2004) *Sold Out: The True Cost of Supermarket Shopping,* London: Vision Paperbacks.

12 Expertise and the calculability of agri-food risks

Richard Le Heron

Border crossing

31 May 2004, Mangere International Airport, Auckland, New Zealand

Incoming international airline passengers surveyed by market research company (for the Ministry of Agriculture and Forestry [MAF]) to ascertain attitudes towards and compliance with New Zealand's biosecurity border regulations.

Questions include probing what travellers knew about biosecurity, degree of compliance on most recent trip, views on extent to which respondents thought compliance should be enforced, and where pre-border information relating to New Zealand biosecurity rules should be made available.

Introduction

This chapter's primary aim is to examine how the rise of 'international competitiveness' has become a key vector in the government of agriculture and food in recent decades and how through the analysis of 'risks' the international competitiveness of a national space called New Zealand comes into existence. A secondary aim is to open up room for contemporary understandings of risk that are more situated. The chapter explores recent experience as a vehicle to ask questions about the emergence of a culture of riskification under neo-liberalizing international agriculture and food relations. This approach highlights both specific and general dimensions of governing competitive participation in the globalizing food economy, as understood from within the context of New Zealand.

I introduce, first, international competitiveness as an integral feature of neo-liberalizing agri-food relations, then risk, as a problematic fact of participating in the globalizing food economy. International competitive pressures spring in part from the creation of the World Trade Organization (WTO) framework, which in turn has propelled the emergence of international discourses of risk, such as biosecurity. As Lupton (1999: 10) contends, the 'high level of anxiety about risk phenomena of all kinds (is) intertwined with uncertainties about sociocultural order.' The genesis of agri-food risks is shown to be tied intimately into the naming of experts, the development of expertise and the normalizing of risk as a mode of thought. These preliminaries enable a discussion of the landscape of agri-food risk in New Zealand in such terms against the background of politically embedded narratives of risk. This introductory discussion forms a context (via the international literature and on-the-ground developments in New Zealand) to an investigation of the changing biosecurity framework in New Zealand and the special role of the Ministry of Agriculture and Forestry (formerly Fisheries) in elaborating the framework. Risk is seen as helping to frame the way agriculture and food are now being imagined, assessed, managed and understood in New Zealand. Finally, I reflect on agri-food risk as 'outcomes of sociocultural processes, serving certain social, cultural and political' functions rather than a 'taken-for-granted objective phenomenon' (Lupton 1999: 2).

The scope and approach followed when investigating the New Zealand experience needs brief comment. In reviewing biosecurity in New Zealand I consider only MAF's role, omitting the biodiversity mandate of the Department of Conservation and the Ministry for the Environment, the similar marine biosecurity mandate of the Ministry of Fisheries, the human well-being concerns of the Ministry of Health and the often overlooked efforts of the Ministry of Foreign Affairs and Trade. The principal source of information on MAF's trajectory of development is the journal *Biosecurity*, produced monthly by MAF to publicize activities and to report New Zealand and overseas technical and legislative developments. This source is supplemented by other official documents. Several sections of the chapter present interpretations that are partly informed by field and archival research published elsewhere (Le Heron 2003a, 2004). With these limitations of method in mind, the preliminary conceptualization and argument outlined in the chapter represents agri-food risks as in-the-making heterogeneous phenomena, embedded in a range of governing frameworks. This is a departure from earlier work in the agri-food field relating to risk, where constitutive processes have usually been neglected.

Context: neo-liberalism, agri-food restructuring and international competition

Agri-food production for export from grassland-based and field-based horticultural ecosystems has been the distinguishing historical anchor of New Zealand's trade-dependent economy. Participating in world trade from a distant economy has continually challenged New Zealand's agri-food producers and governments. This geo–economic relationship was supported by a distinctive regulatory regime, in the form of politically negotiated access for New Zealand producer boards to sell largely undifferentiated commodities to the world. Earnings of foreign exchange from a volume-oriented agricultural sector became central to managing the New Zealand economy (Le Heron 1988). The world crisis in industrialized agriculture (Goodman and Redclift 1989) in the last quarter of the twentieth century, however, induced state-led restructuring of economic and regulatory arrangements in many countries, including New Zealand. New relations of production exposed agri-food producers more directly to the pressures of international competition. Indeed, as Hirst and Thompson (1996: 6) argue more generally, the 'question of the capacity of the economy to produce exports has been transformed into a very different kind of problem, one dominated by the assumption that the only way to avoid being a loser is to become as competitive as possible'. In the context of neo-liberal reforms in export-oriented agriculture, international competitiveness is not simply the surface on which government operates, 'but a means of government: its ties, bonds, force, and affiliations are to be celebrated, encouraged, nurtured, shaped, and instrumentalized in the hope of producing consequences that are desirable for all and for each' (Rose 1996: 335).

Not surprisingly, therefore, we find Hindess (1998: 212), writing from Australia, contending that the 'novel governmental problem (of neo-liberalism) is economic security'. This governmental problem is interpreted here as *risks to achieving international competitiveness, in an environment of trade, production and investment liberalization.* Instead of viewing national agriculture and food as a resource for other aspects of the national community (epitomized in concern over domestic food supply and security), agriculture and food have been seen as *resources for globalizing connections* (Hindess 1998). Such a claim suggests that in the pursuit of international competitiveness in particular contexts, who and what is governed, who governs with what, and what governance actually entails is likely to be in a process of significant revision. Nowhere is this more apparent than in New Zealand's agri-food sector, following the instatement of a neo-liberal agenda in the late 1980s.

Earlier work on the New Zealand scene by agri-food researchers has established two conclusions of immediate relevance, one relating to the contextual dimensions of government interest in risk, and the other relating to how risk is constituted. First, regulation theory and the new political economy of agriculture writings highlight the gravity of the wider economic crisis of the 1970s and 1980s that enveloped the sector (Le Heron 1988, 1993) and the fundamental shifts in economic and political organization behind efforts to realign supply chains to better meet international competition (Campbell and Lawrence 2003; Le Heron and Roche 1999; Le Heron *et al.* 2001). This thinking, on economic and organization transformations consequent upon regulatory change, *contextualizes the 'arrival' of international competitiveness, through the increasing connection of New Zealand to the gathering international momentum of trade liberalization.* The advent of the WTO, the Sanitary and PhytoSanitary (SPS) protocols and Technical Barriers to Trade (TBT) in the 1990s established a geo-political and geo-economic framework which 'invited', if not compelled, New Zealand government attention.

Second, recent applications of neo-Foucaultian and actor network analysis to New Zealand's agriculture (Campbell and Liepins 2001; Larner and Le Heron 2004; Liepins 2000; Liepins and Bradshaw 1999) identify the importance of understanding the constitutive roles of discursive and material practices in shaping new categories for governing. Le Heron (2003a, 2004) extends this thread to explicitly include political projects that are picked up or rejected by sector actors. Political projects are seen as influencing the kinds of objects, practices, subjects and spaces that are being thought into being to displace old categories and the generation of mutual expectations upon which sector actors may rely. Thus, at least when seen through the lens of this internationally informed body of New Zealand agri-food research, the current period is a governmental moment of unprecedented change, distinguished by a broad mix of actors, a mix of political as well as anti-political or technical initiatives, the growth of cultures of calculability, the presence and mobility of new experts and new lines of alignment among activities.

So far this chapter has established the New Zealand agri-food context in which the international literature on risk is being read and has outlined the frameworks being used to undertake this reading. I now review the international risk literature. This sets the scene for later discussion of risks, and the agents and agency of risk at the institutional level.

Constituting agri-food risks

The first part of this section outlines international ideas about risk that are drawn upon by actors to forge a risk lexicon in particular contexts. The second part looks at how actors develop local practices from risk-anticipating encounters in territory.

Knowledge about risk

Fundamental to the conception of agri-food risk are two interrelated perspectives: resource features associated with a territory and connections of a territory with other places. In a liberalizing environment, questions of resource quality pivot more and more on how the resource is valued in the globalizing food economy, while connectivity is tied increasingly into supply-chain relationships. The former includes protection dimensions often labelled biosecurity, the latter focuses on assurances over safety, quality and traceability. Industrialized agriculture regardless of type is underpinned by various technological systems and cultures. According to Van Loon (2002: 204), 'to understand risk in a technological culture, we must look for the elements that disrupt the homogeneity and functional smoothness of the integrated system and its functional domain.' That is, contingencies or unexpected developments encapsulated by the term 'risks', whether internally or externally sourced, become central, since contingencies potentially destabilize production relationships and may threaten capitalist (and other) technological systems. Indeed, O'Malley (2000: 458) suggests that 'risk is imagined, both by the governors who deploy it, and by those who study its deployment, as an element in the conduct of conduct: the effort to align activity with a plan.'

But the nature of risk is neither uniform nor unchanging. Strydom (2002: 12–13) holds that the history of risk discourses, stretching from the 1950s to the present, embraces four phases. 'The first phase was epitomised by the risk assessment debate, the second turned on the question of the comparison and social acceptability of risks, the third saw the emergence of opposition to scientific-technological risks and the concomitant attempt of risk analysts to explain the public perception of risks psychologically. The final phase is characterised by' continuing 'contestation over the social construction of risk in a re-constituted public sphere in which democracy is often invoked against the attempts of scientists, technologists, risk analysts, managers and politicians to restrict the discursive process and thus to displace risks.' In mainstream techno-scientific risk analysis, the concern is the identification of risks, mapping their causal factors, building predictive models of risk relations and people's responses to various types of risk.

This general trajectory was challenged by Beck (1992) who probed the implications of social ordering in terms of risk. In spite of Beck's influence and subsequent writing on the subject,[1] very few studies have attempted critical analyses of agri-food risk. Some exceptions include Enticott (2003), Goven (2003), Hennessy *et al.* (2001), Higgins (2001), Lockie (1998), Morris and Bate (1999), Nerlich (2004), Reilly (2003), Scott *et al.* (2004), and Stassart and Whatmore (2003). In the main, however, these researchers do not couch risk in the manner argued for here. That is, the genesis of agri-food risks, as widely interpreted heterogeneous in-the-making phenomena, is under-explored. This conception requires the explicit addition of the political as well as the governmental.

Where does current practice stand over the framing of the governing of agriculture and food as a problem amenable to risk analysis? In its most elegant form (Phillips and Wolfe 2001; see also Le Heron (2003b) for a review of this book) the global food safety system and the internal biosecurity of nations may be regarded as a set of relations among science, society and trade which can be examined in terms of risk. This follows the Codex Alimentarius Commission's work. In this scheme the focus of science becomes risk assessment, trade that of risk management and safety and the communication of information about risk. This risk triangle informs much governmental activity in the agri-food sphere around the world.

Prior to the examination of New Zealand's recent biosecurity experience it is necessary to establish the central role of experts and expertise in fashioning configurations of risk to meet the priority of international competitiveness and the emergent expectations of agri-food actors in given contexts – all implied in the risk triangle orthodoxy.

Experts, expertise and riskification (or the normalizing of risk)

Explicitly regarding agri-food risk as a problem set in and of technological systems enables an understanding of how governmental actors intervene in the world in an effort to rearrange it according to particular rationalities (Dean 1996, 1999a, b; Rose 1993, 1996, 1999). In contexts such as New Zealand, the conjunction of pressures to meet international competition (met via restructuring) and the evolving discourses of risk (constructed during and out of restructuring) meant new imaginaries of international competition and a new visibility of agriculture and food through risk categories. In the language of governmentality and the actor network approach, a process of 'enframing (ordering, revealing) cultivates particular forms of sense-making ... sense making that find(s) ... logical organisation not from the objects themselves, but from the assemblages in which they emerge ... it organises our being-in-the-world by creating phenomena', to which 'learning to be affected' is expected (Van Loon 2002: 105).

When Dean and Hindess (1998: 9) wrote that 'Problems become known through grids of evaluation and judgement about objects that are far from self-evident', they were alerting us to the constitutive outcomes of getting to know a space. The study by Murdoch and Ward (1997: 307) on the 'statistical manufacture of Britain's national farm' reveals how 'the collection of numbers about various populations allows those populations to be acted upon as they are made increasingly visible and calculable' and amenable to reorganization in line with statistical representations. Crucial to such a development are autonomous actors who 'must ... be equipped with forms of calculation and normalization that both enable and constrain particular forms of behaviour' (Murdoch and Ward 1997: 320). Moreover, their actions should be analysed in terms of 'combinations, associations, relationships, strategies of positioning' to understand the agents' 'calculative agency' (Callon 1998: 12; see also Latour 1987).

The foregoing discussion stressed the socially and contextually constituted nature of risk. We turn now to an examination of the arrival of risk thinking in New Zealand's agri-food sector. In keeping with the argument presented above, attention focuses at first on the wider context, especially the early triggers and responses that began to shape different ideas about priorities relating to New Zealand's agri-food sector. A commitment to contain risks requires expertise and resources to develop the language and systems of risk governance, in particular, territory.

Changing agri-food risk conceptions in New Zealand

New Zealand's neo-liberalizing reforms from the 1980s set in train two relatively autonomous though interdependent trajectories of governmental efforts in the agri-food sector. First, commodity chains began to be transformed into supply chains through realignment of relationships and actors so that supply chains would be a feature of how New Zealand competed internationally and understood as objects of governing in their own right. Second, a new generation of institutions was formed to protect the earning capacities of New Zealand's realigned supply chains, after initial dismantling of the old regulatory apparatus.

New legislation for agriculture and food, passed in the early 1990s, helped situate governmental developments relating to New Zealand agri-food as advantageously as possible for a freer trade world. The 1993 Biosecurity Act, which gave responsibilities to MAF, the 1996 Hazardous Substances and New Organisms Act, which is administered by the Environmental Risk Management Authority, and the Australia and New Zealand Food Authority, gave a template which encouraged the systematic rediscovery of New Zealand's biopotential from the perspective of risk. New Zealand was not alone in going down this path. Food safety scares in the USA in the 1990s triggered change. Much worldwide reform, for instance, stemmed from E.coli 0157:H7 outbreaks in fast-food hamburgers in the USA in 1993 (Juska *et al.* 2000). The USA enacted the Pathogen Reduction Act (1996) based on Hazard Analysis Critical Control Points principles, an approach that is now widely recognized by New Zealand's main trading partners. New Zealand sought to transform the institutional framework for regulating food safety and biosecurity, largely as part of an effort to harmonize with the USA, Canada and the EU, and to conform to WTO directives. Initiatives were nested in the SPS Agreement which 'emphasises the use of international agreed standards' and the SPS rules, which 'were developed around risk analysis, risk management, rigorous science and recognition of the concept of equivalence' (Post Election Briefing 1996: 2). Equivalence is an especially important concept because it focuses on outcomes. Thus for New Zealand this gave independence to develop unique dimensions to its biosecurity capability – and so a potential basis to international competitiveness. The country's intimate ties into the GATT process and more recently that of the WTO ensured New Zealand policy-makers and officials were attuned to the wider governance implications of these processes (e.g. greater cross-border flows, flows of organisms and products derived from organisms, hardware and software needed for movement, intellectual property associated with every aspect).

A pertinent question is: 'In the circumstances of a context like New Zealand, and through concepts available and known in the context, what kinds of risks can be discerned?' For New Zealand, this is a question laden with moral and political dilemmas (Busch 2000). To begin to answer such a question requires consideration of the wider international communities in which New Zealand is a notable player. New Zealand's international prominence in the Genetically Modified Organisms (GMO) and sustainable development (SD) debates places the country under the international spotlight. I argue elsewhere (Le Heron 2004) that in spite of a new order of supply-chain coordination, it is political projects of various kinds that shape the boundaries and content of agriculture and food production.

The past decade has seen New Zealand's agriculture and food entangled in a remarkable contest over what model of agriculture would be officially sanctioned. When scrutinized in terms of the risk-centred discussion of this chapter, the debate over GMO is at heart one about choosing among technological systems of capitalist agri-food production. Whatever systems are favoured, however, the cultural performance of technology is very much about realizing international value from a particular place. Thus for New Zealand the GMO controversy brought to a head the possibility of sharply different landscapes of risk. To consider allowing GMO, against a background of international protest, summed to a political risk management exercise of unparalleled proportions. What resolution of the dilemmas inherent in the GMO controversy has occurred?

In an assumed trade liberal world a regulatory framework for GMOs that *contained the risks* was seen to give New Zealand a trade advantage. From 1988 until 2001, when the Royal Commission on Genetic Modification reported, only field trials of genetically modified plants had been allowed under the umbrella of ERMA. This risk-averse organization could only approve or debar a proposal, and not allow conditional release or development of a new organism. This limited authority reflected more than a century of scientific experience from dealing with the invasion of plants and animals after the onset of organized European settlement. Managing GMOs was something new. A Commission – the traditional means to assemble expertise – was viewed as a way to settle a highly politicized impasse and to chart a new direction. The Commission was 'a world-first comprehensive public examination of GE' (James 2001: C1), involving testimony from around the world as well as an elaborate process on consultation in New Zealand. The catholic collegiate of viewpoints and expertise assembled through the Commission's procedures (over 11,000 submissions, 177 interested parties and 330 expert witnesses) are woven into a text that gradually, but resolutely, moved the weight of judgement towards a twofold finding – manage the risks and preserve opportunities (in research, food and medicine). The Commission's recommendation that a new category, 'conditional release, where the use of a genetically modified organism can be made subject to terms and reporting back, as a further assurance of safety and to enhance the management of risk' (Royal Commission 2001, Executive Summary: 3), portrayed GM as *both thinkable and manageable in New Zealand*. Fundamental to the Commission's argument was the acceptance that New Zealand has a mixed economy, involving four co-existing systems: genetically modified production, conventional land-based production, Integrated Pest Management and an organic component.[2] This mapping, to quote the Commission, holds 'exciting promise, not only for conquering diseases, eliminating pests and contributing to the knowledge economy, but for *enhancing the international competitiveness of the primary industries so important to our country's economic well-being*' (Royal Commission 2001, Executive Summary: 2; emphasis added). A moratorium against conditional releases lapsed in October 2003 (amidst protest). Recent overseas reports suggest the anti-GM mood is being translated into reduced investment in GM research (Collins 2004). Thus the result of the risk assessment was that GMO could be released *into* or developed *in* New Zealand, and once present in New Zealand *could be* contained by risk management techniques. The immediate and longer term problem was regarded as persuading the sceptics. This meant successfully communicating the effectiveness of efforts to limit risk while upholding the virtue of continual vigilance. It also shows that categories of risk are embedded in and derived or modified from political projects as well as from economic and biophysical interactions. Ideas about risk are open to definition and reconstruction through political interpretation.

Biosecurity in the making – and the making of MAF

The previous section established the regulatory and legislative milieu in which MAF sits. We consider now the way MAF as a government agency has been incorporated into and been an active player in shaping the domain of biosecurity.[3] This section explores the idea that biosecurity, in keeping with all risks, is never a stable or fixed object of knowledge. This is done by viewing MAF as a set of internal and external relationships that are aligned increasingly to a dominating purpose – making the biosecurity of New Zealand an uncontestable element of New Zealand's international agri-food competitiveness.

The section starts with an overview of the positioning work achieved discursively to create an isolatable and governable space – New Zealand. 'How then is this space made and kept biosecure?' becomes the challenge. Given the loss of earlier networks as a result of MAF's restructuring in the late 1980s, MAF decided to start again, by systematically assessing risk. This involved a range of surveys, of pest both animal and plant, and of the public, who were perceived as key actors in the surveillance of bio-activity. However, in order to map the landscape of unknown and unverified risks, new levels of expertise were brought to bear on securing the biopolitical worth of New Zealand as a secured agri-food environment. This too required extra initiatives, to obtain and extend access into other countries. The paradoxical tension of connecting to a globalizing food economy, through extensive and world-significant agri-food trade and burgeoning international tourist flows originating from around the globe, presented New Zealand and MAF with a complex biosecurity management goal. Public education was seen as the most effective way to nurture a culture of alertness and cooperation. However, despite world-leading practice in many areas of biosecurity, New Zealand's standing has been regularly called into question by the mobility of 'risk' itself. BSE and Foot and Mouth Disease (FMD) exemplify how biosecurity regimes are exposed to constant threats. The treadmill of continuing improvement in supply-chain management, border management and response management prompted a reassessment of MAF and other organizational structures. Another restructuring currently underway entails a new conception of biosecurity, as a strategic framework for government of bio-activity, embracing a more comprehensive, more holistic, more integrated and more transparent set of rules and procedures.

MAF's claim that activity approximating 'biosecurity is one of the oldest government interventions in New Zealand' (Sherwin 2004: 3; but see Roche (2001) for what was actually meant in early periods) resonates with the Parliamentary Commissioner for the Environment's publication on the subject of biosecurity entitled *New Zealand Under Siege.* MAF's advertised purpose is to create opportunity for and manage risk to New Zealand and the food, fibre and forestry and associated industries, a role derived from the 'importance of trade and tourism to the New Zealand economy and ... the biosecurity risks that accompany these economic mainstays' (O'Neil 2001: 4). In *Guarding Pacific's Triple Star*, the title given to the draft Biosecurity Strategy document released in 2002, MAF represents itself as 'the thin green line', 'frontline biosecurity "police", holding the line against breaches' (Jensen 2001: 1–2). That this posture has contemporary currency (the expanding agency has over 800 of 1,300 staff in biosecurity, with 550 in MAF's Quarantine Service) rests on the interconnections with an increasingly freer trading environment and the constant pressures of international competition that result.

With more open borders, New Zealand has met the world in increasingly diverse ways. Growing tourist numbers by air and ship, dramatically increased import levels including used motor vehicles and cargo arriving in containers, and continued diversification of export markets, have immersed the country deeper into the global circuits of trade and production. To meet the challenges of this widening exposure to the world, MAF reassessed its knowledge base. Four initiatives illustrate this effort. First, MAF's approach to plant biosecurity surveillance has undergone significant development since 1987 when MAF underwent its first major restructure. The restructuring changed the frequency and type of grass-roots contact with the horticultural and arable industries. From 1989 to 1999 MAF undertook an annual survey on specific crops (summerfruit, pipfruit, citrus, cucurbits, tomatoes, grapes, cut flowers, cereals, legumes, stonefruit, subtropicals) to ascertain what pests were associated with the crops in New Zealand in order to be able to develop import health standards with the appropriate associated phytosanitary measures (Stephenson 2001). Second, a pilot survey by AgriQuality New Zealand surveyed 'the animal health information held by a selection of rural veterinary practices and farmers ... [that was] evaluated for its potential to contribute towards MAF's ability to rapidly detect new infectious diseases and emerging trends in disease patterns' (Poland 2001: 5). Information from the computerized records of thirty-nine representative farmers, referred to as 'sentinels' in *Biosecurity*, was transferred to a database for analysis. Two dimensions emerged. In terms of practices, only 14 per cent of visits to veterinarians were for sick animals, and, up to 30 per cent of farms in study areas were not serviced by vets. Moreover, farmers called in vets in a relatively small proportion of animal health problems considered significant by them. Third, MAF conducted a sea container review in 2001 to 2002. Normal practice sees MAF inspect nearly a quarter of landed containers. The survey inspected in detail 13,000 imported sea containers, from which 553 organisms were collected and identified. This new level of knowledge lay behind the ministerial statement, 'New Zealand now has tighter border control measures than any other country in the world' (Sutton 2002: 3). Finally, a series of public surveys were undertaken, with results that concerned MAF. 'It may come as a shock to people close to the subject . . . about half of New Zealanders do not know what "biosecurity" means!' (Sim 2002: 5).

The growth of new knowledge has gone hand-in-hand with increases in personnel and expertise. New Zealanders are regularly involved in international fora connected with biosecurity, such as the Office Internationale des Epizooties, Interim Commission on Phytosanitary Measures and Australian and New Zealand Consultative Group on Biosecurity Cooperation, and contributing to the preparation of risk analysis texts for use in the SPS framework (MacDiarmid 2001). Within New Zealand expertise has now been extended to accredited persons working for recognized agencies able to supervise pre-export activities for live animals and animal germ plasm. New X-ray equipment aids detection at entry. Yet, with disarming honesty, MAF noted in *Biosecurity* that its draft statement on Import Risk Analysis did not originally contain a section on 'Dealing with uncertainty or lack of knowledge'. This is perhaps a signal of a belief in biosecurity as implementation of taken-for-granted 'knowns'.

In some ways the inward orientation of much biosecurity activity means the market access work of MAF is less well known. In many cases a commodity may be prohibited entry to a country because a pest risk has not been conducted, or for other phytosanitary reasons. A 2001 summary of current market access projects published by MAF reveals the extent of such efforts. The projects covered, by country and number, Australia (20), USA (7), Korea (5), Israel (4), South Africa (4), People's Republic of China (3), EU (2), Taiwan (2), Philippines (2), Chile (2), Mexico (1), Uruguay (1), Argentina (1), and Canada (1). The case of capsicum portrays the negotiation of detailed technically justified modifications to existing requirements.

> The USDA has determined that exports of capsicums from New Zealand will need to be from MAF-registered glasshouses insect proofed with self-closing doors and 0.6mm insect mesh on vents. MAF is required to certify that these conditions have been met, including periodic inspections.
>
> (Ogden and Johnston 2001: 10)

The level of expertise has led to the acceptance of an electronic certification system E-cert as a model for use by APEC countries, which:

gives New Zealand a competitive edge because we are one of the few countries in the world that can show what has happened to an exported product from the time the animal or plant enters a factory for processing until the time it gets to the importing country.

(McKenzie 2002: 3)

Prompted by a scorpion scare, where the public took three weeks to notify MAF of the insects, the educational strategy was intensified. The strategy is based on the view that 'Educational activities ... need to begin offshore with travellers and importers, continue at the border where people and goods pass into the country and conclude with a host of practical actions to raise the level of awareness of all New Zealanders about biosecurity' (O'Neil 2001: 3). A novel approach was taken, with the launch of the Protect New Zealand campaign, where the 'Spokesperson' for the programme – an animated beagle called Max – will help deliver the biosecurity message through a TV campaign, publications and brochures as well as the Protect New Zealand website www.protectnz.org.nz (accessed 3 August 2004). Tellingly, 'one of the objectives is to encourage New Zealanders to accept responsibility for actively participating in protecting our country' (Sim 2002: 5). This extends the bounds of biosecurity expertise to all residents!

While seeking public cooperation is vital to responding to much agri-food risk, the external threat to New Zealand's trading image is never far from the surface. Described as 'the ultimate biosecurity threat' (Belton 2001: 3), FMD has a permanent place in New Zealand's biosecurity planning. New Zealand's responses to the progressing BSE scare highlight an ever-changing and delicate politics to New Zealand's position in the globalizing food economy. The baseline is that following confirmation of the further spread of BSE, overseas authorities and consumers sought even greater assurances that BSE countries such as New Zealand were actively looking to the disease and enforcing measures to prevent its spread. As Daly (2001: 11) outlines, the situation is complex:

New Zealand is internationally recognised as being free from BSE. We already meet the World Organisation for Animal Health requirements for BSE freedom. The EU has also recognised New Zealand as BSE free and exempted New Zealand from removing BSE material during processing.

Though encouraging, even reassuring, this is insufficient:

Any further assessments of New Zealand's BSE-free status, will probably require us to show how we comply with the ban on feeding ruminant protein to ruminant animals (a known disease pathway) and with testing of specified categories of cattle for absence of the BSE agent.

(Daly 2001: 11; emphasis added)

Shortly after this commentary MAF announced a Trade Risk Mitigation Survey to ensure 'greater certainty of BSE freedom status', through use of New Zealand's Animal Health Reference Laboratory. On the plant side, Wansbrough and Glover (2002: 5) write: 'Based on the quantity of conventional seed imported and the area of GM crops grown overseas, MAF considers that imports of maize (corn) and oilseed/forage rape (canola) seeds have the greatest chance of being accidentally contaminated with GM varieties.' The worst fears materialized in 2003 when testing procedures for imports failed to detect contamination in imported maize.

Unquestionably, public and political expectations regarding biosecurity have expanded markedly in recent years (Sherwin 2002: 3). MAF has adopted a strategic approach that, rather than setting down a series of recommendations about how biosecurity should be changed, uses an evolving Biosecurity Strategy to provide a series of expectations about what biosecurity should be delivering, both in terms of processes and outcomes. The setting of expectations within a broad legislative framework was felt to lead to a more flexible system and to allow for the evolution of attitudes. In a justification of internal reorganization, a senior MAF officer wrote: 'we need to morph the MAF personality. . . . We are starting to build a culture and support behaviours that are about listening, learning and informing' (Fergusson 2003: 3). The underlying rationale espoused by MAF indicates a different vision of biosecurity and a desire to enrol New Zealanders into the project:

> The main question for MAF was whether it needed to reorganise its biosecurity functions significantly or simply modify its existing sector-focused structure. It decided on the more radical option. The adopted structure is based around two crucial points of intervention – pre-clearance and post-clearance – and an emphasis on cross-system integration.
>
> (Fergusson 2003: 4)

It consists of a pre-clearance directorate spanning risk analysis, import standards, border administration and exports, and a post-clearance directorate ranging over surveillance, incursion investigation and response, pest management and specific incursion response.

Conclusions

This chapter set out to sketch some of the interconnections between agri-food risk and agri-food international competitiveness, on the premise that economic security is attained by being competitive in the globalizing food economy. However, as Hindess (1998: 223–224) perceptively notes, economic security is 'a standing incentive for governmental interference'. New Zealand's experience with new dimensions of food and agricultural governing is consistent with such a view. It is easy to underestimate New Zealand's relatively unusual geo-economic and geo-political position in the globalizing food economy. A situated overview of agri-food risk is required in order to comprehend the New Zealand scene. Boyne's (2003: 106) observation is pertinent: 'It is the cultural context and not the risks themselves that explains whether and how we measure risks and whether we prioritize them.' Therein lies the contribution of this chapter. In spite of the limited nature and basis of this enquiry, it suggests that the frontier of agri-food risk studies needs to move on to examining the conduct of conduct. In New Zealand's case the present aspiration is to keep up with, if not stay ahead of the game, in the risk stakes. A new generation of agri-food risks has been normalized and is being imagined in increasingly sophisticated ways. One immediate outcome has been the re-delineation of 'the border', bracketed by the new concepts of pre-clearance and post-clearance. Thus biosecurity is once more revealed as a contingent process, always becoming. It has a certain shape or content at any given time but is relationally shifting, because it is tied into differing influences and their politics. Getting to know biosecurity is also a way to know the other. In the contemporary New Zealand scene this is about knowing a special face to the global, international competition. Grasping the heterogeneous in-the-making dimensions of MAF's work allows a deeper understanding of agri-food risk and its referent, international competitiveness.

The chapter opened with an arrival interview – a crossing of the border into a space which has been the focus of attention. To the outsider, the range of issues and the degree of probing during the survey may have seemed a little peculiar, even quite invasive, to use the language of biosecurity. The chapter shows how the mentality to pester at the border has grown up from knowing more about the risky world of international competition. It ends with the text going into circulation as part of a book – a mobile vector entering other contexts. The hope is, however, that the chapter's critical appraisal of agri-food risk offers an approach that reveals something of the shadows and unknowns at the edges of a riskification discourse.

Acknowledgements

This research has been supported in part by the Marsden Fund and the Foundation for Research Science and Technology. I also wish to thank Hugh Campbell, Wendy Larner, Nick Lewis and Mike Roche for critical comments on a draft.

Notes

1 The following contain discussion of direct relevance to the argument in this chapter: Adam *et al.* 2000; Belton and Belton 2003; Bessant *et al.* 2003; Boyne 2003; Fox 1998; Hoffman and Oliver-Smith 2002; Kostov and Lingard 2003; Levi 2000; Lupton 1999; Smandych 1999; Tulloch and Lupton 2003; Van Loon 2002.

2 See Campbell and Stuart (Chapter 6, this volume) for discussion on the transformation of the last two categories around the time of the Commission.
3 See Roche (2001) for a discussion of proto-biosecurity in New Zealand in the
   nineteenth and early twentieth centuries.

References

Adam, B., Beck, U. and Van Loon, J. (eds) (2000) *The Risk Society and Beyond: Critical Issues for Social Theory*, London: Sage.

Beck, U. (1992) *Risk Society: Towards a New Modernity*, London: Sage.

Belton, D. (2001) 'Foot and mouth defences strengthened further', *Biosecurity*, 26: 3.

Belton, P. and Belton, T. (eds) (2003) *Food, Science and Society: Exploring the Gap
between Expert Advice and Individual Behaviour,* Berlin: Springer.

Bessant, J., Hil, R. and Watts, R. (2003) *'Discovering' Risk. Social Research and Policy
Making*, New York: Peter Lang.

Boyne, R. (2003) *Risk*, Buckingham: Open University Press.

Busch, L. (2000) 'The moral economy of grades and standards', *Journal of Rural
Studies*, 16, 3: 273–284.

Callon, M. (ed.) (1998) *Laws of the Markets*, Oxford: Blackwell.

Campbell, H. and Lawrence, G. (2003) 'Assessing the neo-liberal experiment in
antipodean agriculture', in R. Almås and G. Lawrence (eds) *Globalization, Localization and Sustainable Livelihoods*, Aldershot: Ashgate.

Campbell, H. and Liepins, R. (2001) 'Naming organics: understanding organic
standards in New Zealand as a discursive field', *Sociologia Ruralis*, 41, 1: 21–39.

Collins, S. (2004) 'Backlash curbs GM investment', *The New Zealand Herald*, 11
June, A7.

Daly, S. (2001) 'New programme reinforces New Zealand's BSE-free status', *Biosecurity*, 31: 10–11.

Dean, M. (1996) 'Putting the technological into government', *History of the Human
Sciences*, 9, 3: 47–68.

Dean, M. (1999a) *Governmentality: Power and Rule in Modern Society*, London: Sage.

Dean, M. (1999b) 'Risks, calculable and incalculable', in D. Lupton (ed.) *Risk and
Sociocultural Theory: New Directions and Perspectives,* Cambridge: Cambridge University Press.

Dean, M. and Hindess, B. (eds) (1998) *Governing Australia: Studies in Contemporary
Rationalities of Government*, Cambridge: Cambridge University Press.

Enticott, G. (2003) 'Risking the rural: nature, morality and the consumption of
unpasteurised milk', *Journal of Rural Studies*, 19, 4: 411–424.

Fergusson, L. (2003) 'Designing system to achieve biosecurity change', *Biosecurity*, 48: 3.

Fox, N. (1998) '"Risks", "hazards" and life choices: reflections on health at work',
Sociology, 32, 4: 665–687.

Goodman, D. and Redclift, M. (eds) (1989) *The International Farm Crisis*, London:
Macmillan.

Goven, J. (2003) 'Deploying the consensus conference in New Zealand: democracy
and de-problematization', *Public Understanding of Science*, 12: 423–440.

Hennessy, D., Roosen, J. and Miranowski, J. (2001) 'Leadership and the provision
of safe food', *American Journal of Agricultural Economics*, 83, 4: 862–876.

Higgins, V. (2001) 'Calculating climate: "advanced liberalism" and the governing
of risk in Australian drought policy', *Journal of Sociology*, 37, 3: 299–316.

Hindess, B. (1998) 'Neo-liberalism and the national economy', in M. Dean and B. Hindess (eds) *Governing Australia: Studies in Contemporary Rationalities of Government*, Cambridge: Cambridge University Press.

Hirst, P. and Thompson, G. (1996) *Globalisation in Question: The Myths of the International Economy and the Possibilities of Governance*, Oxford: Polity Press.

Hoffman, S. and Oliver-Smith, A. (eds) (2002) *Catastrophe and Culture. The Anthropology of Disaster*, Santa Fe: School of American Research Press.

James, C. (2001) 'When life goes on the line', *New Zealand Herald*, 28 July, C1, C3.

Jensen, J. (2001) 'Biosecurity enforcers – holding the line against breaches', *Biosecurity*, 25: 3.

Juska, A., Gouveia, L., Gabriel, J. and Koneck, S. (2000) 'Negotiating bacteriological meat contamination standards in the US: the case of E. coli 0157:H7', *Sociologia Ruralis*, 40: 249–271.

Kostov, P. and Lingard, J. (2003) 'Risk management: a general framework for rural development', *Journal of Rural Studies*, 19, 4: 463–476.

Larner, W. and Le Heron, R. (2004) 'The spaces and practices of a globalising economy', in W. Larner and W. Walters (eds) *Global Governmentality: New Perspectives on International Rule*, London: Routledge.

Latour, B. (1987) *Science in Action: How to Follow Scientists and Engineers through Society*, Cambridge, MA: Harvard University Press.

Le Heron, R. (1988) 'State, economy and crisis in New Zealand in the 1980s: implications for land-based production of a new mode of regulation', *Applied Geography*, 8, 4: 273–290.

Le Heron, R. (1993) *Globalized Agriculture. Political Choice*, Oxford: Pergamon Press.

Le Heron, R. (2003a) 'Creating food futures: reflections on food governance issues in New Zealand's agri-food sector', *Journal of Rural Studies*, 19, 1: 111–125.

Le Heron, R. (2003b) Review of P. Phillips and R. Wolfe (eds) *Governing Food: Science, Safety and Trade*, Montreal and Kingston, Ontario: McGill-Queens University Press, in *Environment and Planning A*, 35, 11: 2087–2088.

Le Heron, R. (2004) 'Re-constituting New Zealand's agri-food chains for international competition', in B. Pritchard and N. Folds (eds) *The Dynamics of Cross-continental Food Commodity Chain Systems*, London: Routledge.

Le Heron, R. and Roche, M. (1999) 'Rapid reregulation, agricultural restructuring and the reimaging of agriculture in New Zealand', *Rural Sociology*, 64, 2: 203–218.

Le Heron, R., Penny, G., Paine, M., Sheath, G., Pedersen, J. and Botha, N. (2001) 'Global supply chains and networking: a critical perspective on learning challenges in the New Zealand dairy and meat commodity chains', *Journal of Economic Geography*, 1: 439–456.

Levi, R. (2000) 'The mutuality of risk and community: the adjudication of community notification statutes', *Economy and Society*, 29, 4: 578–601.

Liepins, R. (2000) 'Exploring rurality through "community": discourses, practices and spaces shaping Australian and New Zealand rural "communities"', *Journal of Rural Studies*, 16: 83–99.

Liepins, R. and Bradshaw, B. (1999) 'Neo-liberal agricultural discourse in New Zealand: economy, culture and politics linked', *Sociologia Ruralis*, 39: 563–582.

Lockie, S. (1998) 'Environmental and social risks, and the construction of "best practice" in Australian agriculture', *Agriculture and Human Values*, 15: 243–252.

Lupton, D. (ed.) (1999) *Risk and Sociocultural Theory*, Cambridge: Cambridge University Press.

MacDiarmid, S. (2001) 'New Zealand experts make major contribution to risk analysis text', *Biosecurity*, 27: 4.

McKenzie, A. (2002) 'Busy times ahead for New Zealand Food Safety Authority', *Biosecurity*, 41: 3.

Morris, J. and Bate, R. (eds) (1999) *Fearing Food: Risk, Health and Environment*, Oxford: Butterworth-Heinemann.

Murdoch, J. and Ward, N. (1997) 'Governmentality and territoriality: the statistical manufacture of Britain's "national farm"', *Political Geography*, 16, 4: 307–324.

Nerlich, B. (2004) 'War on foot and mouth disease in the UK, 2001: towards a cultural understanding of agriculture', *Agriculture and Human Values*, 21: 15–25.

Ogden, S. and Johnson, N. (2001) 'Plant exports market access project', *Biosecurity*, 30: 5.

O'Malley, P. (2000) 'Introduction: configurations of risk', *Economy and Society*, 29, 4: 457–459.

O'Neil, B. (2001) 'Learning to help safeguard New Zealand's biosecurity', *Biosecurity*, 27: 3.

Phillips, P. and Wolfe, R. (eds) (2001) *Governing Food: Science, Safety and Trade*, Montreal and Kingston, Ontario: McGill-Queens University Press.

Poland, R. (2001) 'Vet practices and farms could have key surveillance role', *Biosecurity*, 26: 5.

Post Election Briefing (1996) *1996 Post Election Briefing to the Incoming Minister, Ministry of Agriculture and Forestry*, available at http://www.maf.govt.nz/mafnet/publications/1996-post-election-briefing (accessed 27 July 2004).

Reilly, J. (2003) 'Food risks, public policy and the mass media', in P. Belton and T. Belton (eds) *Food, Science and Society. Exploring the Gap between Expert Advice and Individual Behaviour*, Berlin: Springer.

Roche, M. (2001) 'Biosecurity and proto biosecurity in New Zealand', Paper circulated at the Agrifood 2001 Conference, Palmerston North, New Zealand, December.

Rose, N. (1993) 'Government, authority and expertise in advanced liberalism', *Economy and Society*, 22, 3: 327–356.

Rose, N. (1996) 'Governing "advanced" liberal democracies', in A. Barry, T. Osborne and N. Rose (eds) *Foucault and Political Reason: Liberalism, Neo-liberalism and Rationalities of Government*, London: University College of London Press.

Rose, N. (1999) *Powers of Freedom: Reframing Political Thought*, Cambridge: Cambridge University Press.

Royal Commission on Genetic Modification (2001) *Report of the Royal Commission on Genetic Modification*, Wellington: Department of Internal Affairs.

Scott, A., Christies, M. and Midmore, P. (2004) 'Impact of the 2001 foot-and-mouth disease outbreak in Britain: implications for rural studies', *Journal of Rural Studies*, 20, 1: 1–14.

Sherwin, M. (2002) 'Biosecurity effort to step up another gear', *Biosecurity*, 40: 3.

Sherwin, M. (2004) 'Leadership of new biophysical agency taking shape', *Biosecurity*, 53: 3.

Sim, K. (2002) 'What do New Zealanders really know about biosecurity?', *Biosecurity*, 34: 5.

Smandych, R. (ed.) (1999) *Governable Places. Readings on Governmentality and Crime Control*, Dartmouth: Ashgate.

Stassart, P. and Whatmore, S. (2003) 'Metabolising risk: food scares and the un/remaking of Belgian beef', *Environment and Planning A*, 35, 3: 449–462.

Stephenson, B. (2001) 'Changing times and plant pest surveillance', *Biosecurity,* 27: 16–17.

Strydom, P. (2002) *Risk, Environment and Society,* Oxford: Oxford University Press.

Sutton, J. (2002) 'Huge debt owed to biosecurity frontliners', *Biosecurity,* 33: 3.

Tulloch, T. and Lupton, D. (2003) *Risk and Everyday Life,* London: Sage.

Van Loon, J. (2002) *Risk and Technological Change: Towards a Sociology of Virulence,* London: Routledge.

Wansbrough, D. and Glover, G. (2002) 'Assuring imported seeds are not GM', *Biosecurity,* 33: 5.

Index